DOT-VNTSC-FAA-07-04

Air Traffic Organization Operations Planning
Human Factors Research and Engineering Group
Washington, DC 20591

**Electronic Flight Bag (EFB):
2007 Industry Review**

**Michelle Yeh
Divya C. Chandra**

U.S. Department of Transportation

Research and Innovative Technology Administration

John A. Volpe National Transportation Systems Center

Cambridge, MA 02142

April 2007

This document is available to the public through the National Technical Information Service, Springfield, Virginia, 22161

Notice

This document is disseminated under the sponsorship of the Department of Transportation in the interest of information exchange. The United States Government assumes no liability for its contents or use thereof.

Notice

The United States Government does not endorse products or manufacturers. Trade or manufacturers' names appear herein solely because they are considered essential to the objective of this report.

			Form Approved
REPORT DOCUMENTATION PAGE			OMB No. 0704-0188

Public reporting burden for this collection of information is estimated to average 1 hour per response, including the time for reviewing instructions, searching existing data sources, gathering and maintaining the data needed, and completing and reviewing the collection of information. Send comments regarding this burden estimate or any other aspect of this collection of information, including suggestions for reducing this burden, to Washington Headquarters Services, Directorate for Information Operations and Reports, 1215 Jefferson Davis Highway, Suite 1204, Arlington, VA 22202-4302, and to the Office of Management and Budget, Paperwork Reduction Project (0704-0188), Washington, DC 20503.

1. AGENCY USE ONLY (Leave blank)	2. REPORT DATE April 2007	3. REPORT TYPE AND DATES COVERED Final Report, April 2007
4. TITLE AND SUBTITLE Electronic Flight Bag (EFB): 2007 Industry Review		5. FUNDING NUMBERS DD304/FA6Y
6. AUTHOR(S) Michelle Yeh and Divya C. Chandra		
7. PERFORMING ORGANIZATION NAME(S) AND ADDRESS(ES) U.S. Department of Transportation John A. Volpe National Transportation Systems Center Research and Innovative Technology Administration Cambridge, MA 02142-1093		8. PERFORMING ORGANIZATION REPORT NUMBER DOT-VNTSC-FAA-07-04
9. SPONSORING/MONITORING AGENCY NAME(S) AND ADDRESS(ES) U.S. Department of Transportation Federal Aviation Administration Air Traffic Organization Operations Planning Human Factors Research and Engineering Group 800 Independence Avenue, SW Washington, D.C. 20591 Program Manager: Dr. Tom McCloy		10. SPONSORING/MONITORING AGENCY REPORT NUMBER
11. SUPPLEMENTARY NOTES		
12a. DISTRIBUTION/AVAILABILITY STATEMENT		12b. DISTRIBUTION CODE

13. ABSTRACT (Maximum 200 words)

This document, which is based on information from March 2007, provides an overview of Electronic Flight Bag (EFB) systems and capabilities, with particular focus on the systems' human interface. It updates the April 2005 EFB Industry Review (Yeh and Chandra, 2005). The information in this document will be useful to anyone interested in the EFB market, including the Federal Aviation Administration (FAA), customers, operators, manufacturers, and researchers.

The report is divided into three sections. The first section briefly reviews EFB research conducted by the Volpe Center over the past several years and the results of that research. The second section describes products and services offered by several system providers and integrators. The third section is a list of software providers. A list of references, including policy and research documents is provided at the end of this report.

14. SUBJECT TERMS Electronic Flight Bag, EFB, industry review, flight deck technology, avionics, manufacturers, EFB systems			15. NUMBER OF PAGES 59
			16. PRICE CODE
17. SECURITY CLASSIFICATION OF REPORT Unclassified	18. SECURITY CLASSIFICATION OF THIS PAGE Unclassified	19. SECURITY CLASSIFICATION OF ABSTRACT Unclassified	20. LIMITATION OF ABSTRACT

NSN 7540-01-280-5500

Standard Form 298 (Rev. 2-89)
Prescribed by ANSI Std. 239-18
298-102

METRIC/ENGLISH CONVERSION FACTORS

ENGLISH TO METRIC

LENGTH (APPROXIMATE)
- 1 inch (in) = 2.5 centimeters (cm)
- 1 foot (ft) = 30 centimeters (cm)
- 1 yard (yd) = 0.9 meter (m)
- 1 mile (mi) = 1.6 kilometers (km)

AREA (APPROXIMATE)
- 1 square inch (sq in, in^2) = 6.5 square centimeters (cm^2)
- 1 square foot (sq ft, ft^2) = 0.09 square meter (m^2)
- 1 square yard (sq yd, yd^2) = 0.8 square meter (m^2)
- 1 square mile (sq mi, mi^2) = 2.6 square kilometers (km^2)
- 1 acre = 0.4 hectare (he) = 4,000 square meters (m^2)

MASS - WEIGHT (APPROXIMATE)
- 1 ounce (oz) = 28 grams (gm)
- 1 pound (lb) = 0.45 kilogram (kg)
- 1 short ton = 2,000 pounds (lb) = 0.9 tonne (t)

VOLUME (APPROXIMATE)
- 1 teaspoon (tsp) = 5 milliliters (ml)
- 1 tablespoon (tbsp) = 15 milliliters (ml)
- 1 fluid ounce (fl oz) = 30 milliliters (ml)
- 1 cup (c) = 0.24 liter (l)
- 1 pint (pt) = 0.47 liter (l)
- 1 quart (qt) = 0.96 liter (l)
- 1 gallon (gal) = 3.8 liters (l)
- 1 cubic foot (cu ft, ft^3) = 0.03 cubic meter (m^3)
- 1 cubic yard (cu yd, yd^3) = 0.76 cubic meter (m^3)

TEMPERATURE (EXACT)
[(x-32)(5/9)] °F = y °C

METRIC TO ENGLISH

LENGTH (APPROXIMATE)
- 1 millimeter (mm) = 0.04 inch (in)
- 1 centimeter (cm) = 0.4 inch (in)
- 1 meter (m) = 3.3 feet (ft)
- 1 meter (m) = 1.1 yards (yd)
- 1 kilometer (km) = 0.6 mile (mi)

AREA (APPROXIMATE)
- 1 square centimeter (cm^2) = 0.16 square inch (sq in, in^2)
- 1 square meter (m^2) = 1.2 square yards (sq yd, yd^2)
- 1 square kilometer (km^2) = 0.4 square mile (sq mi, mi^2)
- 10,000 square meters (m^2) = 1 hectare (ha) = 2.5 acres

MASS - WEIGHT (APPROXIMATE)
- 1 gram (gm) = 0.036 ounce (oz)
- 1 kilogram (kg) = 2.2 pounds (lb)
- 1 tonne (t) = 1,000 kilograms (kg) = 1.1 short tons

VOLUME (APPROXIMATE)
- 1 milliliter (ml) = 0.03 fluid ounce (fl oz)
- 1 liter (l) = 2.1 pints (pt)
- 1 liter (l) = 1.06 quarts (qt)
- 1 liter (l) = 0.26 gallon (gal)
- 1 cubic meter (m^3) = 36 cubic feet (cu ft, ft^3)
- 1 cubic meter (m^3) = 1.3 cubic yards (cu yd, yd^3)

TEMPERATURE (EXACT)
[(9/5) y + 32] °C = x °F

QUICK INCH - CENTIMETER LENGTH CONVERSION

QUICK FAHRENHEIT - CELSIUS TEMPERATURE CONVERSION

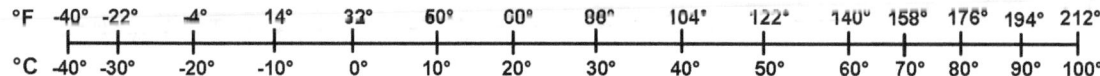

For more exact and or other conversion factors, see NIST Miscellaneous Publication 286, Units of Weights and Measures. Price $2.50 SD Catalog No. C13 10286   Updated 6/17/98

Preface

This report was prepared by the Human Factors Division of the Office of Aviation Programs at the John A. Volpe National Transportation Systems Center (Volpe Center). It was completed with funding from the Federal Aviation Administration's (FAA) Human Factors Research and Engineering Group (AJP-61) in support of the Aircraft Certification Service Avionics Branch (AIR-130). We would like to thank our FAA program manager, Dr. Tom McCloy, as well as our technical sponsor, Colleen Donovan for providing suggestions and feedback. Many thanks also to the many manufacturers who generously provided information for the industry review. As with any system development, changes in the design occur frequently; as a result, this information is only accurate for a short period of time. For each product, the manufacturer's website is provided where more up to date information may be found.

The views expressed herein are those of the authors and do not necessarily reflect the views of the Volpe National Transportation Systems Center, the Research and Innovative Technology Administration, or the United States Department of Transportation.

Feedback on this document can be sent to Michelle Yeh (Michelle.Yeh@volpe.dot.gov) or Divya Chandra (Divya.Chandra@volpe.dot.gov). Further information on this research effort can be found at www.volpe.dot.gov/hf/aviation/efb.

Table of Contents

Executive Summary iii
Acronyms iv
Introduction 1
Overview of Volpe Center EFB Research 2
System Providers 3
    ADR 5
    Airbus 7
    Aircraft Management Technologies (AMT) 9
    AirGator, Inc 11
    ARINC 13
    Astronautics 15
    CMC Electronics Inc. 17
    DAC International 19
    Flight Deck Resources (FDR) 21
    FlightPrep, Stenbock and Everson, Inc. 23
    Goodrich Sensors and Integrated Systems 25
    Jeppesen 27
    navAero 29
    OBDS.com (On-Board Data Systems) 31
    Paperless Cockpit 33
    Rockwell Collins 35
    Teledyne Controls 37
    Universal Avionics 39
    Virtual Papyrus Inc. 41
Software Providers 43
References 45
Trademark Notices 47
Websites 48

Executive Summary

This document provides an overview of Electronic Flight Bag (EFB) capabilities, capturing the state of the EFB industry as of March 2007. It will be useful to anyone interested in the EFB market, including the Federal Aviation Administration (FAA), customers, operators, manufacturers, and researchers. This document is an update to the April 2005 Volpe Center EFB Industry Review (Yeh and Chandra, 2005).

The document consists of three main sections. The first section contains a brief review of EFB research conducted by the Volpe Center over the past several years. The second section contains descriptions of products from EFB system providers and integrators. The material in this section was gathered through collaboration with industry and from information provided at demonstrations and in websites, brochures, and trade journal reports. A general picture of the product lines is provided, with an emphasis on the user interface. Information on FAA approvals received or in progress is also included. The third section lists EFB software providers. References to both policy and research documents are included at the end of this document.

Acronyms

AC	Advisory Circular
ACARS	Airborne Communications Addressing and Reporting System
ACO	Aircraft Certification Office
ADS-B	Automatic Dependent Surveillance-Broadcast
AEG	FAA Flight Standards Aircraft Evaluation Group
AIM	Aeronautical Information Manual
AIR	FAA Office of Aircraft Certification
AMLCD	Active Matrix Liquid Crystal Display
AWIN	Aviation Weather Information
CD	Compact Disk
CDL	Configuration Deviation Lists
CDTI	Cockpit Display of Traffic Information
CDMA	Code Division Multiple Access
CE	Conformité Européene (European Compliance)
COTS	Commercial-off-the-shelf
CPDLC	Controller Pilot Data Link Communications
CPU	Central Processing Unit
CRJ	Canadair Regional Jet
CWIN	Cockpit Weather Information
DoD	Department of Defense
DVD	Digital Versatile Disk
EASA	European Aviation Safety Agency
ECL	Electronic Checklists
EFB	Electronic Flight Bag
EMI	Electromagnetic Interference
EVS	Enhanced Vision System
eTAWS	Early Terrain Awareness Warning System
FAA	Federal Aviation Administration
FMS	Flight Management System
FSB	Flight Standardization Board
FSDO	Flight Standards District Office
GPRS	General Packet Radio Services
GSM	Global Systems Mobile
GUI	Graphical User Interface
HDU	Hard Disc Unit
HITS	Highway-in-the-sky
IFR	Instrument Flight Rules

LAN	Local Area Network
LCD	Liquid Crystal Display
LED	Light-Emitting Diode
LRU	Least-Recently Used
LVDS	Low Voltage Differential Signaling
MEL	Minimum Equipment List
METAR	Aviation Routine Weather Report
NACO	National Aeronautical Charting Office
NEXRAD	Next Generation Weather
ND	Navigation Display
NOTAM	Notices to Airmen
NTSC	National Television System(s) Committee
OEM	Original Equipment Manufacturer
OIS	Onboard Information System
OS	Operating System
PC	Personal Computer
PCMCIA	Personal Computer Memory Card International Association
PDA	Personal Digital Assistant
PDF	Portable Digital Format
PID	Pilot Information Display
RPU	Remote Processor Unit
SATCOM	Satellite Communications
SDK	Software Development Kit
SOP	Standard Operating Procedure
SST	Solid State Tablet
STC	Supplemental Type Certificate
SVGA	Super Video Graphics Adapter/Array
TAWS	Terrain Awareness System
TC	Type Certificate
TCAS	Traffic Alert Collision Avoidance System
TCP/IP	Transmission Control Protocol Internet Protocol
TFR	Temporary Flight Restriction
TFT	Thin-Film Transistor (screens)
TSO	Technical Standard Order
TWLU	Terminal Wireless LAN Unit
UMTS	Universal Mobile Telecommunications System
US	United States
USB	Universal Serial Bus
VHF	Very High Frequency

WAC	World Aeronautical Chart
WiFi	Wireless Fidelity
WLAN	Wireless Local Area Network
WSVGA	Wide Super Video Graphics Adapter/Array
XGA	Extended Graphics Adapter/Array (1024x768 resolution)
XML	Extensible Markup Language

Introduction

The Electronic Flight Bag (EFB) industry continues to grow and mature. EFB manufacturers offer systems for a wide range of operations, including commercial air transport, corporate, and general aviation. This EFB industry review provides an overview of EFB systems capabilities, capturing the state of the EFB industry as of March 2007. This report is provided to the Federal Aviation Administration (FAA) but the information in this report is intended to be of use to anyone interested in the EFB market. This review is an update of two previous reviews (see Appendix A of Chandra, Yeh, Riley, and Mangold (2003) and Yeh and Chandra (2005)).

This review is divided into three sections. The first section contains a brief review of EFB research conducted by the John A. Volpe National Transportation Systems Center (Volpe Center) over the past several years. This information may be useful to those who are new to the topic, or for those who need a quick review of all the different resources available.

The second section, the bulk of this report, contains two-page entries describing the offerings from each of several providers and integrators. Many manufacturers now offer product lines with multiple hardware and software solutions configured into various packages. In this report, a general picture of the product lines is provided rather than details for specific EFB models. This review highlights aspects of the EFB interface (e.g., the display size, the number and types of controls, and the available applications) rather than hardware aspects (e.g., memory, processor speed). Information on FAA approvals received or in progress is also included. The entries for each system also indicate whether it runs custom software (i.e., developed in-house) or could run a third party commercial-off-the-shelf (COTS) application. This information is a rough indicator of the level of integration across the EFB hardware and various software applications.

The last section of this review contains a list of software providers. These applications can be installed on EFB hardware that is purchased separately. Detailed information is not provided for the individual software applications.

A list of reference documents, including policy and research documents, is provided at the end of this document. The Volpe Center research reports in the reference list are available at www.volpe.dot.gov/hf/aviation/efb. FAA policy documents are available at www.faa.gov, under the Regulations and Policies section. In addition, completed Flight Standardization Board (FSB) reports for particular EFB models, which are available to the public at www.opspecs.com, are also listed.

Overview of Volpe Center EFB Research

The Volpe Center has been working with the EFB industry since 1999, when the EFB research program began. Industry interaction has always been a key component of this program. The Volpe Center connects EFB industry stakeholders, including regulators, customers, manufacturers, and end users (pilots). Volpe Center research informs the FAA of industry trends through periodic industry reviews, and it informs industry about FAA-sponsored efforts that may help them in their EFB endeavors at industry meetings and events. Many of the Volpe Center EFB research efforts are well summarized in a report by the Flight Safety Foundation (2005).

When this research began, the goal was to identify EFB human factors issues in order to improve knowledge, awareness, and understanding of EFB human factors considerations by all parties. Working with the Air Transport Association Digital Display Working Group, the Volpe Center produced a key report in 2000, known as the "Version 1 EFB human factors considerations document" (Chandra and Mangold, 2000). This report was referenced in the March 2003 FAA Advisory Circular (AC) 120-76A, *Guidelines for the certification, airworthiness, and operational approval of electronic flight bag computing devices*.

In September 2003, the Volpe Center released an updated "Version 2" report on EFB human factors considerations (Chandra, et al., 2003), which supersedes the Version 1 report. The Version 2 document is still current as of 2007, but may be updated in the future. As in the Version 1 report, the Version 2 document provides guidance on the general EFB system human factors considerations, and it covers four applications in detail: Electronic Documents, Electronic Checklists, Flight Performance Calculations, and Electronic Charts. Although the document is long, it is easy to browse, and has proven to be an especially useful reference for system developers.

The next phase of the Volpe Center EFB research was directed at streamlining and standardizing FAA evaluations of EFBs. With the aid of the EFB industry, the Volpe Center constructed five tools that could be used by FAA or industry evaluators to assess the human factors issues associated with particular EFBs. The tools were designed for use by evaluators who were not necessarily human factors experts to provide useful feedback, even with a relatively small investment in time.

Chandra and Yeh (2006) contains descriptions of the five Volpe Center EFB assessment tools and practical information on when and how to use each tool. The tools are included as appendices to that report. Additionally, four of the Volpe Center EFB assessment tools are included as appendices in FAA Notice 8200.98. This Notice allows the FAA inspector to use the tools, or not, at his/her discretion. These tools may also be of use to EFB manufacturers and customers, who could use the tools to improve their system design or to anticipate the results of a regulatory evaluation. Details describing the Volpe Center's effort to develop the first set of tools, for use by the FAA Aircraft Certification Service, are in Chandra, Yeh, and Riley (2004); in particular, three appendices, E (Update to Industry on Tools for Structured Evaluations of EFB Usability), F (Generic Task List for Evaluations), and G (Sample Observer Notes and Evaluation Feedback), of that document may be particularly useful for EFB developers who would like to understand how to use the tools for in-house evaluations. The biggest benefit of using these tools is that their early use can reduce the redesign associated with poor system interfaces, and ensure that the EFB system is more usable in the long run.

System Providers

Nineteen system providers participated in this review. The term *system providers* is used here to refer to companies that offer an integrated hardware and software solution.

The Volpe Center worked collaboratively with a representative from each of the companies included in this section. The final entry was limited to two pages, and was approved by the company for publication in this review.

The following information was gathered from each system provider:

- Product name.
- Website(s) where more information can be found. The text in the tables is hyperlinked to the manufacturer's site. A list of websites is included at the end of this document.
- Any FAA approvals received or in progress. When available, the FAA office issuing the approval is noted. A list of publicly available FSB reports is provided in the References section.
- Environmental Qualifications (if any): A list of environmental tests conducted or in progress (e.g., RTCA DO-160E). Note that Class 1 and 2 systems do not require compliance with RTCA-DO160E per FAA Advisory Circular (AC) 120-76A.
- A brief (100-word) overview of the product.
- The intended hardware class (i.e., Class 1, 2 or 3 system).
- Description of the hardware user interface. Specifically, the size of the display, special brightness characteristics of the display (e.g., whether the display been enhanced for use in bright light or night flight), communications capabilities, type of controls (and in particular whether the system provides a touch screen, stylus, buttons, mouse/cursor control, and/or keyboard), mounting and/or stowage solutions, hardware style, and accessories that could enhance the use of the EFB.
- Applications supported. System providers were asked to indicate whether the software is developed in-house or a COTS application, i.e., software provided by a third-party manufacturer that the EFB provider could install. Custom applications for charts, checklists and flight performance calculations were classified based on their functionality into the categories listed in Table 1 below. A list of the range of applications offered and software providers is available in the Software Providers section.

Note that EFB solutions vary in architecture, and these differences are reflected in the entries for each company. Some EFBs run applications from several software providers, e.g., a charting application from one manufacturer, an electronic document viewer from another, and a weather application from a third. The EFB operating system could be a standard system (e.g., Microsoft Windows or Linux), or it could be a custom-written operating system. The applications and operating system may all have a different look and feel and vary in their user interface. Some system providers offer a custom operating system that organizes the various third party applications so that the pilot can interact with all EFB applications using a standardized interface. Others go a step further and work with a customer to create a complete, integrated EFB solution, offering services ranging from integration of the EFB with the aircraft, EFB approvals, and after-market support. Finally, some manufacturers place more of an emphasis on software offerings, providing a customized software solution for use on *any* hardware system. The entries for these manufacturers provide more detailed information on their software applications, and forgo the information pertaining to the hardware user interface.

Electronic Charts	• **Raster**: A digital image created by scanning a paper chart into an electronic format. Portions of the chart could be scanned separately, allowing limited user control over what information is displayed. When zooming in or out of a chart, all display elements on the chart (including text) grow larger or smaller, respectively. If the chart is rotated, all objects will rotate correspondingly (i.e., the text will not remain upright). • **Vector-based**: Display elements are stored as individual objects. When zooming in or out of the display, lines and symbols are redrawn, whereas text may keep its same size and orientation.
Electronic Checklists	• **Viewer only**: The system only displays the checklist items. The operator is responsible for entering and maintaining the checklist items. The system does not indicate which checklist item is active, and does not monitor checklist completion. • **Error checking**: The system tracks the checklist item status and alerts the pilot under various conditions. For example, the system tracks whether all items in the checklist have been completed and alerts the pilot accordingly. • **Automated error-checking**: The checklist system is aware of aircraft status and tracks item status automatically. • **Active checklist**: The pilot can control aircraft status through checklist actions. For example, if the pilot selects the item "turn on engine," the engine would be turned on.

Electronic Documents	1. Documents are classified as one of the following:
	   • **Viewer Only**: The look of a paper document is preserved with few or no additional features. The length of a line of text in the document is independent of the size of the display, so increasing the text size may cause the document to run off the sides of the screen.
	   • **Viewer with Additional Features**: The look of the paper document is preserved, as in the Viewer Only document. In addition, text links are used throughout so the pilot can see information on a specific topic by clicking a word, and keyword search is supported.
	   • **Markup Language**: Combines text and extra information about the structure or presentation of the text. Additional features are supported, e.g., automatic word wrapping as the font size changes and more sophisticated search functionality, such as search across documents for related entries.
	2. Electronic documents that support a note-taking feature are indicated.

Flight Performance Calculations	1. A list of the functions supported (e.g., Weight & Balance, Takeoff and Landing Performance) 2. Provides details regarding whether the calculations are based on data for that specific individual aircraft (i.e., with a specific tail number) or whether the data are for a generic aircraft of that make and model.

Table 1. Functional Classification of Software Applications.

ADR		Location:	Rochester Hills, MI & West Palm Beach, FL
Product(s)	Flight Guide (FG)-1610, FG-6000, FG-7000TL, FG-8000		
Website(s)	• www.adrsoft.com • Product Information: FG-1610, FG-6000, FG-7000, FG-8000		
FAA Approvals	• FG-3600: STC for Boeing 747-400 (Los Angeles ACO) • FG-5000 and FG-6000: STC's for Embraer ERJ-135 (Dallas ACO), Challenger 601/604 (Chicago ACO) • FG-7000T: STC Effort on Citation 550/560, Project Number ST-85895SC-T (Dallas ACO) • FG-8000: STC Pending for various Bombardier aircraft (Chicago ACO)		
Environmental Qualifications	• FG-6000: Decompression • FG-7000: Electromagnetic Interference (EMI), Decompression • FG-8000: RTCA DO-160E Qualified		

Product Overview

The FG-Series EFB systems comprise of commercial-off-the-shelf (Fujitsu) touch-screen computers, customized for use in unique aviation environmental conditions. In particular, the displays have been specially enhanced to improve screen readability in sunlight, and some systems offer a "night flight" dimming feature that allows the brightness of the screen to be adjusted to levels that will not compromise night vision. The FG-8000 remote screen is a custom designed product with extensive RTCA DO-160E testing used as both a class 2 and 3 device. The FG-Series EFB systems are compatible with Jeppesen's family of products.

FG-1610   *FG-7000*

Photos courtesy of ADR.

Hardware Class(es)	Class 1, 2, and 3 systems
Display Size	8.4" SVGA (800 x 600) or 8.9" WSVGA (1024 x 600), WXGA (1280 X 768)
Brightness	Some systems offer *high-bright* screens that offer enhanced brightness for viewing in all lighting conditions and night-flight dimming control.
Communications	Varies by system: 802.11 A/B/G wireless

ADR		Location:	Rochester Hills, MI & West Palm Beach, FL
Controls	<ul><li>Touch screen. Stylus or finger entry accepted</li><li>Buttons vary by system. Some have membrane buttons for numeric keypad, zoom functionality, and 4-button cursor control. FG-8000 accepts input from membrane input keys. Software switches only on Fujitsu-based products</li><li>Laptop/slate tablet systems offer keyboard</li></ul>		
Mounting and/or Stowage	Customizable: Cockpit- or yoke-mountable using RAM or gooseneck mounting solutions, kneeboard with legstrap or STC mounting cradles and arms		
Hardware Style	Laptop/slate tablet and pen-tablet computers (Fujitsu) Custom designed RTCA DO-160E tested units for class 2 and 3 systems		
Accessories	USB GPS receiver, docking station, dual serial PCMCIA card, USB keyboards, serial adapter, 429 to serial (RS-232) converter, compact flash dual serial adapter, USB composite video adapter, power adapter, battery/battery charger, extra stylus, screen protectors		
Applications Supported			
Operating System	<ul><li>Microsoft Windows (XP Tablet Edition, Windows 2000 Pro, or XP Pro)</li><li>Linux</li><li>Flight Command System (FCS) Software Interface</li></ul>		
Electronic Charts	✓ COTS application (Jepp View)		
Electronic Checklists	✓ Error checking		
Electronic Documents	✓ Non-interactive		
Flight Performance Calculations	✓ COTS application		
Flight Planning	✓ COTS application		
Video Surveillance	✓ COTS application		
Weather	✓ COTS application		
Other	Notetaker (application allows users to write notes in their own handwriting and save them)		

Airbus	Location: Toulouse, France
Product(s)	FlySmart with Airbus EFB Class 3 for Type A/B applications
Website(s)	N/A
FAA Approvals	The hardware is certified by the Joint Aviation Authorities (JAA) on the A330/340. Applications are approved by the Joint Operations Evaluation Board (JOEB) of the European Aviation Safety Agency (EASA).
Environmental Qualifications	Certification is compliant with RTCA-DO160 E.

Product Overview

The Airbus EFB solution is based on two displays, which are called the Onboard Information Terminal (OIT) displays. Each is located on sliding tablet and linked to two computer processing units (CPUs) installed on the aircraft floor. The two CPUs are connected together via Ethernet link (network exchange) and video link (video exchange). The CPUs are avionics certified computers that can be connected to Airbus Aircraft Network and communications devices. Several functions, such as video exchanges and cross check, are included to allow airlines to go up to no-paper operations.

The Airbus EFB supports both Airbus applications and third party (Type A) applications. Airline-specific Type A applications can be integrated into the main menu.

Photo courtesy of Airbus.

Hardware Class(es)	Installed and certified installation with Type A/B applications. No Type C applications.
Display Size	12" diagonal, 1024 x 768
Brightness	OIT displays are certified for night conditions (<0.7 nits) as well as in sunny conditions.
Communications	Airbus EFB is part of the Airbus Aircraft Network when it is linked to the Aircraft Server with access to aircraft wireless and Satcom. Airbus EFB can connect by Ethernet to specific communications devices.
Controls	OIT displays are controlled with touch screen and/or wired keyboard storable in lateral stowage. All applications are designed for touch-screen. OIT hardware interface buttons on display: • Mouse/cursor control, Video Left/Right (L/R) exchanges, CPU L/R selection per display • Brightness dimming • Reset USB port on display. USB and Ethernet port on lateral sides (L/R) of the cockpit Keyboard port on display.
Mounting and/or Stowage	OIT displays on long range family (A330/340) are installed on the sliding tablet CPU are installed on the aircraft floor for the serial solution
Hardware Style	Teledyne provides OIT displays and CPUs. OIT and CPUs are linked with video buses (Ethernet or optical fibers). CPUs can be connected to four different video inputs. OIT CPUs offer USB and Ethernet ports.
Accessories	All software can be installed on laptop for Class1 use

Airbus	Location: Toulouse, France
Applications Supported	
Operating System	Microsoft Windows XP pro. A main menu has links to avionics parameters (flight number, tail number, flight plan). A data loader and configuration manager are also supported.
Electronic Charts	✓ On request. Vector based electronic-charts from third party suppliers can be hosted.
Electronic Checklists	Provided on Airbus flight deck displays. ECL management on EFBs is not needed
Electronic Documents	✓ Airbus electronic documentation ship set is supported in HTML and XML format. WML documents planned for 2008: FCOM, MEL,CDL. Flight briefing document allow pilots to load free format text or pictures, update these documents during flight, and download after flight using a USB key.
Flight Performance Calculations	✓ Airbus Less Paper Cockpit (LPC) applications include weight and balance, MEL, FCOM, and performance modules (takeoff and landing performance, cruise performance planned in 2008). These applications can be used on all Fly-by-Wire aircraft families (A320 family & A340 family).
Flight Planning	Flight briefing documents folder (free text /picture format) and communication manager
Video Surveillance	Optional. Can be connected on OIT video ports
Weather	Under study.
Other	Technical Logbook/Connection to Airbus Aircraft Network for electronic maintenance operations Third-party airport moving maps can be hosted.

Aircraft Management Technologies (AMT)		Location: Dublin, Ireland
Product(s)	Flightman Applications Suite	
Website(s)	www.flightman.com	
FAA Approvals	N/A	
Environmental Qualifications	N/A	

Product Overview

The Flightman Applications Suite delivers a hardware-independent EFB solution for any Class 1, 2, or 3 device. AMT is an approved supplier to Boeing, Rockwell Collins, and European Space Agency. The suite of tools is capable of communicating over a wide variety of channels including USB, Cellular, WiFi, Satcom and ACARS.

The following configurable applications are provided:

- **Electronic Journey Log**: an electronic version of existing paper forms carried on a commercial aircraft. It includes hours and cycles, crew information, fuel management, delays and landing information, and vendor services used.
- **Electronic Technical Log** (with separate flight crew and maintenance user versions): an electronic version of current paper-based techlog held aboard an aircraft that allows an airline to manage defects within its fleet. Flightman Technical Log is configurable to any aircraft type and provides a standard user interface across a mixed fleet.
- **Performance calculations** for takeoff and landing
- **Weight and Balance**
- **Large Content Management** (electronic documentation management & presentation): Ground based document/content management system enabling the remote distribution and management of content to other EFBs using Flightman software. This content can include:
    o Charts
    o Aircraft manuals
    o Company documents and circulars
    o Flight plan
   Publish/subscribe infrastructure allows data to broadcast across the entire fleet or to specific groups defined by tail number, aircraft type, fleet type, user, etc.
- **Cabin Surveillance**
- **Passenger Relationship Management**

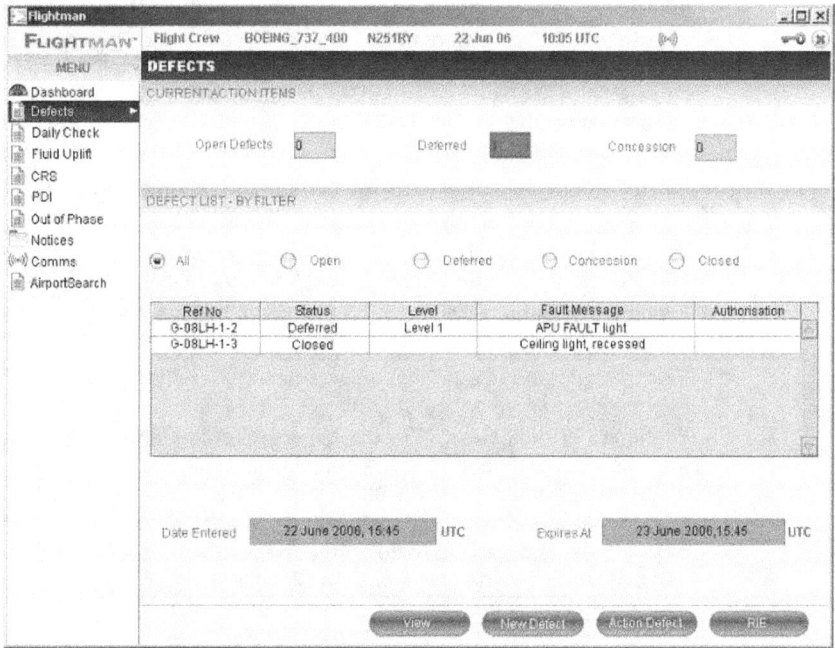

Flightman Technical Log

| Aircraft Management Technologies (AMT) | Location: Dublin, Ireland |

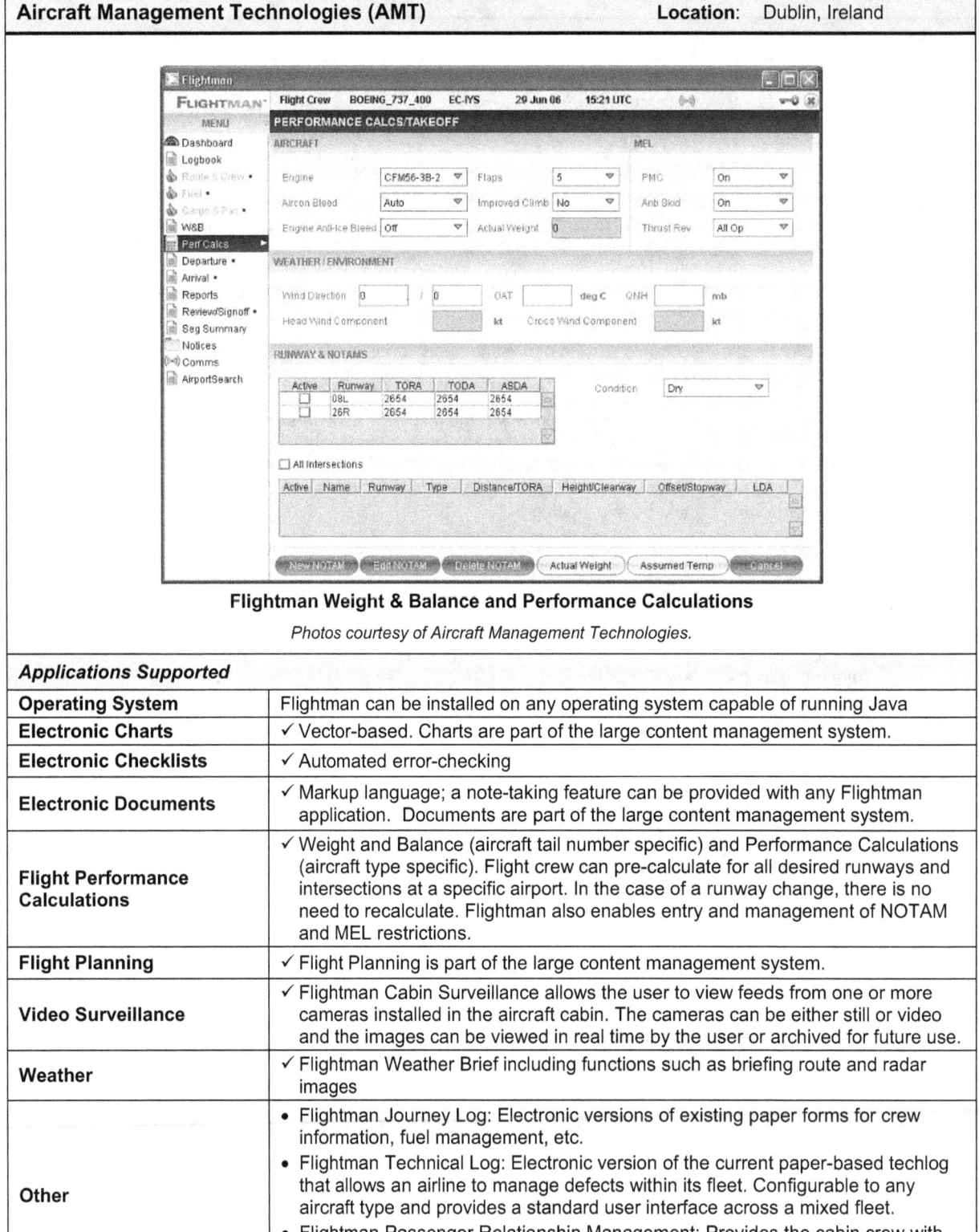

Flightman Weight & Balance and Performance Calculations

Photos courtesy of Aircraft Management Technologies.

Applications Supported	
Operating System	Flightman can be installed on any operating system capable of running Java
Electronic Charts	✓ Vector-based. Charts are part of the large content management system.
Electronic Checklists	✓ Automated error-checking
Electronic Documents	✓ Markup language; a note-taking feature can be provided with any Flightman application. Documents are part of the large content management system.
Flight Performance Calculations	✓ Weight and Balance (aircraft tail number specific) and Performance Calculations (aircraft type specific). Flight crew can pre-calculate for all desired runways and intersections at a specific airport. In the case of a runway change, there is no need to recalculate. Flightman also enables entry and management of NOTAM and MEL restrictions.
Flight Planning	✓ Flight Planning is part of the large content management system.
Video Surveillance	✓ Flightman Cabin Surveillance allows the user to view feeds from one or more cameras installed in the aircraft cabin. The cameras can be either still or video and the images can be viewed in real time by the user or archived for future use.
Weather	✓ Flightman Weather Brief including functions such as briefing route and radar images
Other	• Flightman Journey Log: Electronic versions of existing paper forms for crew information, fuel management, etc. • Flightman Technical Log: Electronic version of the current paper-based techlog that allows an airline to manage defects within its fleet. Configurable to any aircraft type and provides a standard user interface across a mixed fleet. • Flightman Passenger Relationship Management: Provides the cabin crew with passenger information to better serve the needs of each customer. The system can be installed in the galley or on a Class 1 device.

AirGator, Inc	Location: Mt Kisco, NY
Product(s)	• NAVPad (30 Gb Hard Disk Unit (HDU), 8 - 24 Gb Solid State Tablet (SST)) • NAVPad 10 (30 Gb HDU, 8 - 24 Gb SST) • NAVPadQ1 (30/60 Gb HDU, 32 Gb SST) • NAVPad Flip convertible notebook/tablet (30/60 Gb HDU)
Website(s)	www.airgator.com
FAA Approvals	AirGator's NAVPads qualify as Class 1 and Class 2 EFBs per AC120-76A
Environmental Qualifications	Some NAVPad units have undergone DoD testing and certification and passed 4' drop, humidity, temperature and vibration tests.
Product Overview	
AirGator's NAVPad EFB solutions combine NAVAirEFB moving map software with real-time satellite broadcast weather capability and NAVAir Approaches electronic approach charts viewer on a wide range of aviation-optimized tablet computers. NAVPad systems are available with high capacity Solid State FLASH as well as G-Shock-protected hard drive storage. Any Windows compatible software (such as Jeppesen NAVSuite) can also be used on any NAVAir device. The software is also offered separately and can be used with any Microsoft Windows-based system.	

Photos courtesy of AirGator, Inc.

Hardware Class(es)	Class 1 and 2
Display Size	NAVPad (HDU and SST): 8.4" 800 x 600, 1.7 lb. NAVPad 10 (HDU and SST): 10.4" 1024 x 768 2.1 lb NAVPad Q (HDU and SST): 7" 800 x 480 1.4 lb NAVPad Flip (HDU): 8.9" 1280 x 768 2.1 lb

AirGator, Inc	Location: Mt Kisco, NY
Brightness	All units feature daylight readable touch screens, dimmable for night flying NAVPad: Outdoor sunlight readable dimmable to 0 nits in 3% steps NAVPad Q: 300 nits dimmable to 100 nits NAVPad Flip: Transflective sunlight readable dimmable to 0 nits in 6% steps
Communications	All units feature Ethernet, WiFi 802.11g wireless and Bluetooth connectivity. Broadband EVDO, EDGE, 3G and other options available
Controls	Controls layout varies by system. • All displays have touch screen and are stylus-free • Buttons. Systems generally have four hard keys, 5-way navigation button, and power button • Hardware buttons for brightness/dimming, screen rotation and onscreen keyboard control • Keyboard: wireless or USB keyboard stowed separately
Mounting and/or Stowage	Portable devices. Can be mounted in a variety of ways including yoke mounts, center console mount, knee pads and other options.
Hardware Style	Finger touch screen tablet computers
Accessories	GPS units, memory upgrades, antennas, USB CD/DVD drive, weather receivers, keyboards, power supplies, battery/backup battery, extra stylus, screen protectors
Applications Supported	
Operating System	Microsoft Windows XP Tablet, Professional, Embedded Editions
Electronic Charts	✓ Vector Based (NAVAir Moving Map, NAVAir Approaches Approach), Chart Viewer (COTS applications, such as Jeppesen NavSuite), and Raster Based (COTS applications) charting options available.
Electronic Checklists	✓ COTS application
Electronic Documents	✓ Document viewer and note taking capability including onscreen clearance writer and scratchpad.
Flight Performance Calculations	✓ COTS applications
Flight Planning	✓ NAVAirEFB or COTS applications
Video Surveillance	✓
Weather	✓ NAVAirEFB XM WX and WSI/Sirius viewer with moving map. Compatible with other viewers.
Other	Aircraft logbooks, Standard Operating Procedures (SOPs), squawk lists, communications and dispatch, email, other.

ARINC	Location: Annapolis, MD
Product(s)	eFLYBook
Website(s)	• www.arinc.com • Product Information: eFLYBook
Approvals	None at this time. Will work with customers as needed.
Environmental Qualifications	None at this time. Plan to begin DO-160E testing.

Product Overview

The ARINC eFLYBook electronic document (eDoc) viewer is based on a major development in display and document viewing technology called electronic paper, which distinguishes it from other EFBs. The device is lightweight (under 1 pound) and has an 8.1-inch display that is highly readable even in direct sunlight. The device is just 1/2-inch thick and has a battery life of 20 hours of continuous use.

The eFLYBook contains all U.S. digitized terminal procedures, IFR high and low altitude enroute charts, U.S. airport facility directory, the Code of Federal Regulations/Aeronautical Information Manual (AIM), and can accommodate other user-installed electronic documents, in PDF or XML format (e.g., manuals, checklists, forms, books, etc.)

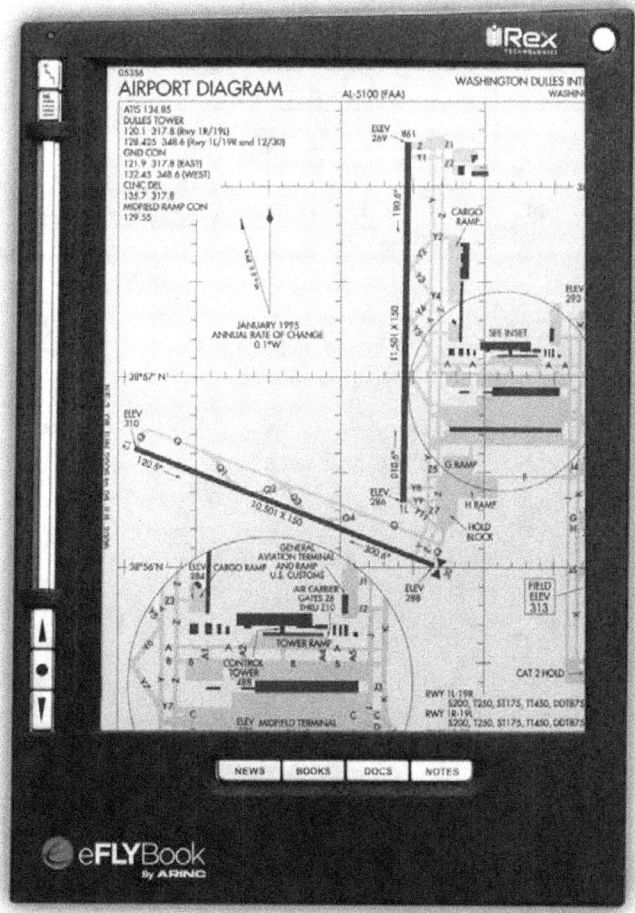

Photo courtesy of ARINC.

Hardware Class(es)	eFLYBook is intended to be a Class 1 or 2 system
Display Size	8.1" electronic paper, 768 x 1024, 16 levels of gray
Brightness	The system is readable in direct sunlight

ARINC	Location: Annapolis, MD
Communications	WIFI 802.11g wireless LAN, 10/100MB wired LAN
Controls	<ul><li>Touch screen</li><li>Stylus: Active, stowed in back of device.</li><li>Buttons: 10 hard keys (power, 4 menu buttons, 3 scroll buttons) and a scroll bar. Hard keys or stylus can be used to interact with the application.</li></ul>
Mounting and/or Stowage	Kneeboard and yolk mounting -style
Hardware Style	eReader (electronic reader with a paper-like display)
Accessories	Carrying case, aircraft power adapter, RAM mount, extra stylus, stylus tether
Applications Supported	
Operating System	Linux
Electronic Charts	✓ (Raster charts with zoom and pan functionality)
Electronic Checklists	✓ (Interactive; the pilot can use the stylus to mark whether an item has been completed but the checklist application does not provide error checking nor is it active. Checklist can be saved for future reference by the pilot or others)
Electronic Documents	✓ (Non-Interactive, but provides stylus-based data-input that allows pilots to add handwritten notes)
Flight Performance Calculations	In development
Flight Planning	
Video Surveillance	
Weather	In development (Text only)
Other	Notices to Airmen (NOTAMs), many other aviation publications (e.g., Code of Federal Regulations, AIM etc.)

Astronautics	Location: Milwaukee, WI
Product(s)	Class 2 and 3 Pilot Information Display (PID). The Class 3 is a standard option on Boeing production aircraft, and upgraded version is standard on 787.
Website(s)	<ul><li>www.astronautics.com</li><li>EFB effort: PID</li></ul>
FAA Approvals	<ul><li>Certified (Tailored Linux) and Uncertified (Windows) applications for Class 3 PID that has been TC'ed and STC'ed on Boeing commercial and military aircraft</li><li>Certified (Linux) or Uncertified (Windows) applications for Class 2 PID</li><li>Class 3 Level C, D and E software supported (DO-178B). Certification completed. Class 2 can support certified or uncertified software</li></ul>
Environmental Qualifications	Class 2 and 3 are RTCA DO-160D Compliant Hardware

Product Overview

The Astronautics EFBs are avionic quality displays with adaptable hardware and software configuration. These EFBs consist of two displays, installed on either side of each pilot's seat. Single- or dual-processor options are available. In the dual-processor design, one processor is configured to run the Linux operating system and the other Microsoft Windows, allowing certified and non-certified applications to be isolated. The single-processor design can be configured to support either certified or uncertified applications. The hardware is compliant with RTCA DO-160D for use in all phases of flight, and is backed by Astronautics worldwide support organization.

 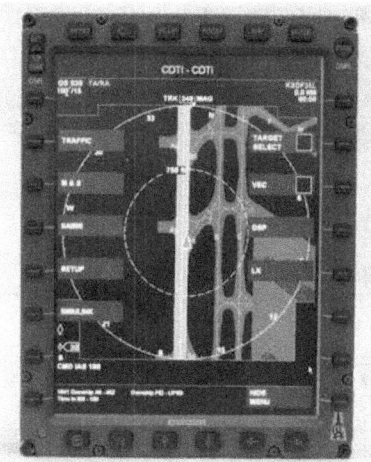 

Photos courtesy of Astronautics

Hardware Class(es)	Class 2 and 3 systems are avionics quality equipment that minimizes COTS obsolescence, and provides a cockpit quality device, backed by Astronautics and its worldwide support organization.
Display Size	Class 2/3 provides a common 10.4", XGA resolution (1024 x 768) AMLCD
Brightness	High-contrast display with light-emitting diode (LED) backlighting with a wide range of brightness from sunlight readable to dark cockpit operations. Further, the luminescence is compatible with other equipment in the cockpit.
Communications	Class 3 ARINC-42 (16Rx), RS-422/485, Ethernet, USB & PS/2, and TCP/IP Video, Fiber optic link between Display and Electronic Unit, and interfaces with the Terminal Wireless LAN Unit (TWLU). The Class 3 also supports two way interfaces with on-board sensors. Class 2 ARINC-429 (x4Rx), RS-422/232, Ethernet, USB & PS/2, Video, 802.11 wireless

Astronautics	Location: Milwaukee, WI
Controls	<ul><li>Touch Screen</li><li>Display controls: 16 programmable bezel soft keys, 12 dedicated function keys, brightness increase/decrease key, and power control.</li><li>Systems operate independently but provide cross talk capability enabling each crewmember to view the other's data and to support the cross loading of databases.</li><li>Virtual keyboard and on-display cursor capability</li><li>Can support an external keyboard.</li></ul>
Mounting and/or Stowage	Fixed or adjustable. One display for each pilot; adjustable arm, window mount, or fixed
Hardware Style	EFBs are avionic quality hardware consisting of two display units and single or dual processor units.<ul><li>The Class 3 units are connected by a fiber optic cable, and can be 100 feet apart. The Class 3 Electronic Unit is installed in the EE bay.</li><li>The Class 2 Electronic Unit can be installed in any area within 15 feet of the Display Unit.</li></ul>
Accessories	The Class 2/3 systems do not require accessories, but support the communications interface described above.
Applications Supported	
Operating System	Class 3: two independent processor/ hard drives. One hosts Linux for certified (Type C) applications, second hosts Windows 2000/XP for uncertified applications. Dual hard drives provide 160 GB of mass storage. Class 2: single processor/ hard drive hosts Windows XP and provides 80 GB or optionally 160 GB of mass storage. Hosting Linux operating system (OS) is an option. The Class 2/3 EFBs are open architecture systems that can host any application selected by the operator, and supports the customer developing/customizing his own applications.
Electronic Charts	✓ (COTS)
Electronic Checklists	✓ (COTS)
Electronic Documents	✓ (COTS)
Flight Performance Calculations	✓ (COTS)
Flight Planning	✓ (COTS)
Video Surveillance	✓ (COTS)
Weather	✓ (COTS)
Other	CDTI application (Astronautics provides Type C certified application for ADS-B based, ACSS SafeRoute program that provides sequence & merging of traffic, and surface surveillance.) ADS-B/TCAS, runway incursion prevention, terrain avoidance, eTAWS, countermeasure display, FLIP charts, Falcon View, performance weight & balance, maintenance, data link (SATCOM, GateLink, Link 16, etc.), CPDLC, Combat Track II, and others.

CMC Electronics Inc.	Location: Montreal, Quebec
Product(s)	PilotView Class 2 EFB
Website(s)	<ul><li>www.cmcelectronics.ca</li><li>Product Information: PilotView EFB</li></ul>
FAA Approvals	<ul><li>Compliant to Class 2 requirements identified in FAA AC120-67a and EASA TGL-36</li><li>FAA and European Aviation Safety Agency (EASA) STC Approvals:<ul><li>Gulfstream (G100/150, GII/GIIB, GIII, GIV, GV, GVSP)</li><li>Dassault (Falcon 10, 50, 900, 2000)</li><li>Bombardier (Global 5000, Global Express, CL604)</li><li>Pilatus PC-12</li><li>Boeing 737, 767</li></ul></li></ul>
Environmental Qualifications	Qualified to RTCA DO-160E standards
Product Overview	
The PilotView EFB consists of a self-contained electronic display and processor unit (EDU) installed via an integrated mount with a companion power and interface expansion module (EEMU). The EFB hosts a suite of custom software and COTS applications on its CMC-controlled Microsoft Windows operating system and custom software utility to support business jet operators and the air transport market.	

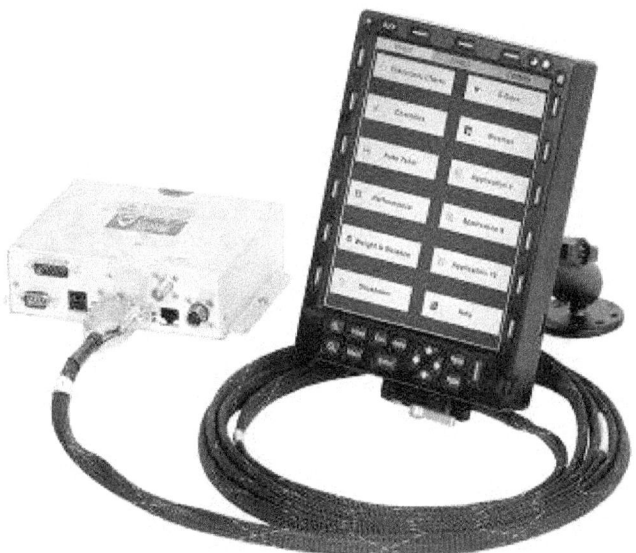

Photo courtesy of CMC Electronics.

Hardware Class(es)	Class 2
Display Size	8.4" active matrix XGA (1024 x 768) display
Brightness	Fully dimmable; enhanced display for sun-light and night readability (1-800 nits)
Communications	Integrated Wireless 802.11a/b/g, PCMCIA, USB 2.0, Ethernet, ARINC 429, RS232, RS422, Video, Discrete inputs/outputs
Controls	<ul><li>Film on Glass resistive touch screen</li><li>15 soft 'line select' function keys; dedicated keys for zooming, dim, bright, video and application control; 4-button joystick, page up, page down</li><li>Unique integrated slide-in, FMS-style keyboard</li></ul>
Mounting and/or Stowage	Yoke- or side-mounted display configurations. Separate processor unit mounted using single connector that interfaces with aircraft power; removable via integrated latching mount.

CMC Electronics Inc.	Location: Montreal, Quebec
Hardware Style	Custom built integrated display and processor
Accessories	Power Supply Adapter for EDU ground use and EDU carrying case are supplied as accessories.
Applications Supported	
Operating System	Microsoft Windows (XP Pro) & custom embedded OS
Electronic Charts	✓ (COTS – Jeppesen, LH Systems eRouteManual, EAG Aerad)
Electronic Checklists	✓ (Error checking, PDF style or PDF-based checklist application)
Electronic Documents	✓ (e-DOC browser – Viewer with Additional Features and will support markup language)
Flight Performance Calculations	✓ (Weight & Balance; takeoff and Landing calculations, which incorporate information from airport obstacle databases, airport NOTAMs, and weather. Computations based on individual aircraft type)
Flight Planning	
Video Surveillance	✓ (Optional video viewer that can be combined with CMC SureSight Enhanced Vision System (EVS) camera systems and partners multi-camera video systems)
Weather	✓ (COTS - available package with WSI, Honeywell (WINN) and WxWorx weather products)
Other	Customizable Journey Log, EVS head down display, Note Taker, SideView (pilot collaborative tool enabling pilots to share views of their current active applications)

DAC International	Location: Austin, TX
Product(s)	GEN-X EFB
Website(s)	www.dacint.com
FAA Approvals	Currently STC applications for CRJ (all types; Los Angeles ACO), DC-8 (Dallas ACO), B727 (Dallas ACO), B757/767 (Chicago ACO), and L328 (Atlanta ACO)
Environmental Qualifications	Qualified to RTCA DO-160E standards
Product Overview	
DAC's GENESYS solution including the GEN-X EFB provides airlines a complete and total system solution for the paperless cockpit. The system includes EFB hardware, EFB software, electronic chart data, Gatelink, server software, and a cabin surveillance system. Each component of the system is designed to get the most from the rest of the interconnected components.	

Photo courtesy of DAC.

Hardware Class(es)	Class 2 or 3
Display Size	8", 10", or 12"
Brightness	900+ Nits to <1 nit transflective screen for viewing in bright sunlight
Communications	ARINC 429, Ethernet, RS-232, USB 2nd USB on Display Unit
Controls	Touch screen display with display on-off button and dimmer controls
Mounting and/or Stowage	Installed system, RPU (Remote Processor Unit) can be installed in the cockpit (Class 2) or in the E&E bay (Class 3). Display has mounting holes tapped for direct mount attachment in the rear.
Hardware Style	Two separate units: Display Unit and RPU
Accessories	Pre built umbilical cable for display to RPU interface (no maximum length restrictions).

DAC International	Location: Austin, TX
Applications Supported	
Operating System	Microsoft Windows (XP Professional) plus GENESYS Application Manager/Shell software which prohibits operators (other than Administrator) from seeing windows. All functions are controlled through the shell which also supports a means to automatically update company data, chart data, and software.
Electronic Charts	✓ COTS application – choice of MapTech, Jeppesen or Lido. Unique Clipboard user interface for Origin, Destination, Enroute and Alternate
Electronic Checklists	✓ Viewer only for initial release, active checklist will be an option
Electronic Documents	✓ Viewer with Additional Features: The look of the paper document is preserved. In addition, hyperlinks are used throughout, and keyword search is supported as well as go to page number.
Flight Performance Calculations	✓ COTS application
Flight Planning	✓ COTS application
Video Surveillance	✓ Supported with video server
Weather	✓ XMWX WxWorx supported
Other	Third party applications, e.g., for synthetic vision, surface moving map, electronic logbooks & more.

Flight Deck Resources (FDR)	Location: Irvine, CA
Product(s)	SkyTab 770HB, SkyTab 900R, SkyTab 1100, SkyTab 2200, and full Software Suite
Website(s)	<ul><li>www.flightdeck.aero</li><li>Product Information: SkyTab 770HB, SkyTab 900R, SkyTab 1100</li></ul>
FAA Approvals	The SkyTab 1000NG is certified and approved for use on a Boeing 767 aircraft as a Class 2 EFB system. Pending MD11, DC10 and Dash 8 STCs for the Sky Tab 1100 EFB.
Environmental Qualifications	Qualified to RTCA DO-160D

Product Overview

Flight Deck Resources provides integrated hardware and software Class 1 and 2 EFB systems. The SkyTab product line consists of ruggedized touch-screen computers with sunlight-readable displays. Applications supported include Flight Deck Resources custom software for displaying charts and documents as well as COTS software (e.g., JeppView, WxWorx, WSI InFlight). Integrates with ARINC 429, weather, Iridium Satellite, etc.

 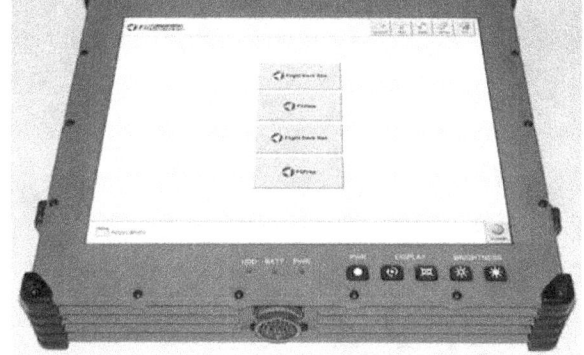

SkyTab 1100 with Approach Plate   *SkyTab 1100 showing FliControl*

Photos courtesy of Flight Deck Resources.

Hardware Class(es)	Class 1 and 2
Display Size	8.4" SVGA (800x600) or 10.4" XGA (1024 x 768)
Brightness	All systems allow for both hardware and software based dimming from 2 to 800nits.
Communications	Varies by system: wireless 802.11b, Ethernet, RS232, ARINC 429
Controls	Varies by system<ul><li>All are touch screen</li><li>Stylus is not needed, finger touch used on all systems.</li><li>Buttons are for screen on/off, bright/dim, and unit power.</li><li>Mouse/cursor control can be added with USB based accessories.</li><li>Keyboard is on screen or USB based</li></ul>
Mounting and/or Stowage	Articulating mounting arms usable during all phases of flight for Class 2 devices as well as kneeboard or stowage options for the SkyTab 770HB. FDR provides for complete certified mounting solutions.
Hardware Style	Tablet computers
Accessories	SkyTab 1100 powers off ship power(24 VDC); SkyTab 770HB has power adapter, battery pack, external USB CD/DVD drive, docking station

Applications Supported

Operating System	Microsoft Windows (XP Professional); some systems offer Linux
Electronic Charts	✓ COTS windows-based applications (e.g., Jeppesen, NavTech EAG/eRM, Lido)
Electronic Checklists	✓ Error checking (FliPrep). Can also host COTS applications

Flight Deck Resources (FDR)	Location: Irvine, CA
Electronic Documents	✓ Markup Language (FliView)
Flight Performance Calculations	✓ COTS application
Flight Planning	✓ COTS application
Video Surveillance	✓ EFB can accept and display video
Weather	✓ COTS application
Other	FliControl – user interface that serves as a "control panel" for other applications

FlightPrep, Stenbock and Everson, Inc.		Location: Aurora, Oregon
Product(s)	ChartCase Professional	
Website(s)	• www.flightprep.com • Product Information: ChartCase, EFB Systems	
FAA Approvals	Complies with FAA AC 120-76A as a Class 1 or Class 2 EFB software system.	

Product Overview

ChartCase Professional provides electronic charting, XM Weather, and advanced flight planning functions, delivering paperless cockpit capabilities for most Windows-based computers. All Sectional Charts, WAC Charts, High/Low Enroute Charts, Instrument Procedures, Airport Diagrams, and vector charts for the U.S. are provided. Additionally, in-cockpit weather capability is supported using the WxWorx receiver so that NEXRAD, METARs, TAFs, TFRs, and more can be overlaid for the route and flight path. A Synthetic Vision (Highway in the sky (HITS)) feature displays flight information in 3D format and can show supplemental GPS based flight telemetry. A Terrain Awareness Function (TAWS) is also supported.

ChartCase Professional

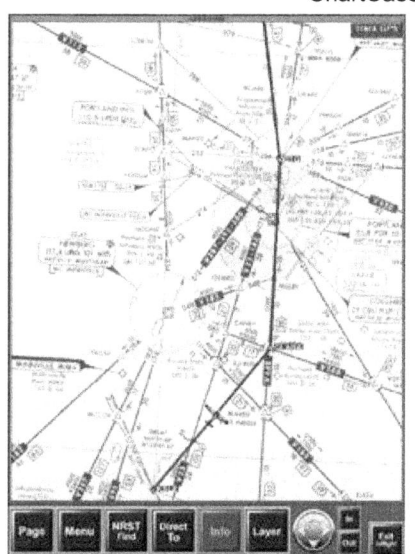
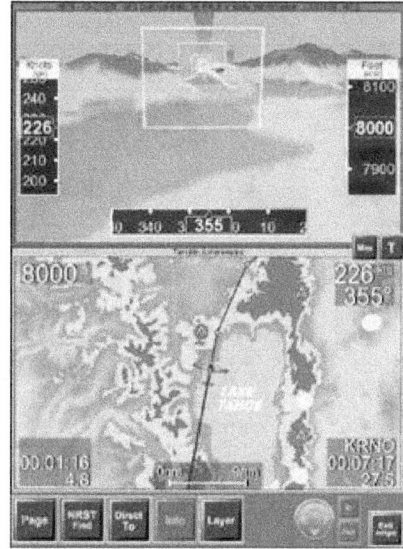
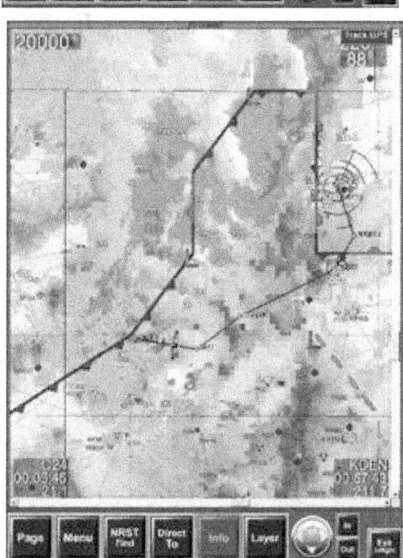
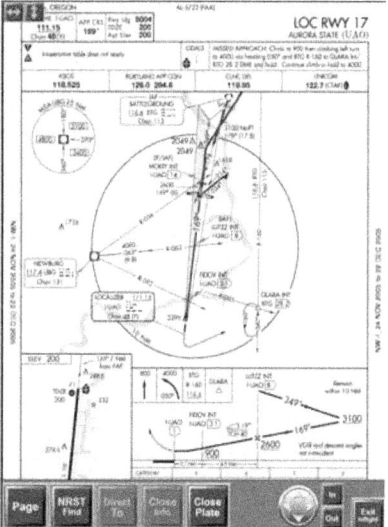

Photos courtesy of FlightPrep.

Hardware Class(es)	Class 1 and 2 EFB Devices

FlightPrep, Stenbock and Everson, Inc.	**Location:** Aurora, Oregon
Display Size	From 7" WVGA (640 X 480) to 60" monitor and beyond
Brightness	Some systems offer enhanced displays for viewing in direct sunlight
Communications	Varies by system: 801.11 b/g wireless, Ethernet, Bluetooth
Controls	Varies by system. • All have either touch screen or digitizer pen functionality. • Tablet computers have a minimum of five hard-keys that perform tablet PC Functions • Stylus: Active on Motion Computing tablet PCs; inactive on others. Tetherable Stylus, Stowage location available in casing of EFB • EFBs function with any type of USB or wireless USB or Bluetooth Keyboard
Mounting and/or Stowage	Portable, kneeboard, yoke or other types of mounting devices to attach EFB on a temporary (Class 2) installation
Hardware Style	Tablet computer, Slate Computers, Laptops, Desktops, Software works on most Windows based PC's
Accessories	Bluetooth GPS receiver, weather receiver, power adapters, weather antennas
Applications Supported	
Operating System	Microsoft Windows XP and Microsoft Windows Vista compliant
Electronic Charts	✓ Raster and Vector-based charts
Electronic Checklists	✓ Error checking: Integrated checklist functionality with completion status buttons for each item available
Electronic Documents	✓ Viewer only; provides creating, viewing, printing, PDF functionality
Flight Performance Calculations	✓ Calculations are based on data for that specific individual aircraft. User may input performance data for numerous aircraft. Program comes with generic data for over 20 aircraft that is modifiable by user Functions supported: Weight & balance, climb/descent performance, fuel planning, altitude analysis tool for selecting efficient cruising altitude based upon forecasted winds
Flight Planning	✓ Tools for routing, filing, and weather provided for off-line or on-line use.
Video Surveillance	✓ COTS application for tablet computers
Weather	✓ XM Weather support when used with Bluetooth or USB XM Radio Receiver
Other	Data Updates Available at 3 different frequencies from FlightPrep: • 28 Day Data Subscriptions (IFR Current Update) • 6 Times a year data subscription • 3 Times a year data subscription

Goodrich Sensors and Integrated Systems	Location: Burnsville, MN
Product(s)	Goodrich Electronic Flight Bag
Website(s)	<ul><li>www.goodrich.com</li><li>Additional Providers: Business Jet Aircraft Completions (BJAC; www.bjacaerospace.com)</li><li>Product Information: EFB</li></ul>
FAA Approvals	<ul><li>STCs for:<ul><li>Bombardier Global Express (in progress through Transport Canada), Global 5000, and Challenger 604 (in progress)</li><li>Boeing 737NG (in progress)</li></ul></li></ul>
Environmental Qualifications	Qualified to RTCA DO-160E standards

Product Overview

The Goodrich Electronic Flight Bag consists of two adjustable display modules located on either side of each pilot, and two computer modules which can be installed almost anywhere in the aircraft. The system has an open design to integrate with various communication protocols including ARINC 429, Ethernet, USB, SATCOM, wireless, and GSM/UMTS. The hardware hosts any Windows-based software developed by Goodrich or other third party manufacturers.

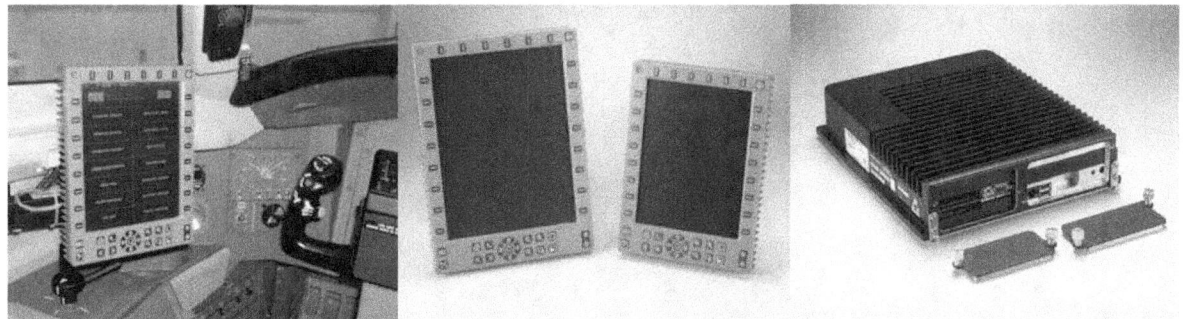

Photos courtesy of Goodrich Sensors and Integrated Systems.

Hardware Class(es)	Class 2 and Class 3
Display Size	8.4" or 10.4", 1024 x 768
Brightness	Enhanced for readability in all lighting conditions; brightness/contrast control plus day/night mode; 170° viewing angle; luminance levels to 750 nits for the 10.4" display and 550 nits for the 8.4" display.
Communications	ARINC 429, Ethernet, MIL-1553, ACARS, 802.11 a/b/g wireless, GSM/UMTS
Controls	<ul><li>Touch Screen</li><li>Programmable Buttons. 22 assignable bezel keys, 4-button joystick with center, navigation and zoom buttons, brightness contrast, video transfer, and real time screen rotation key. Inputs can be through the touch screen or bezel keys depending on software configuration.</li></ul>
Mounting and/or Stowage	The Goodrich EFB system can be configured as installed equipment or mounted depending on the specific aircraft and customer requirements. The display module is mounted in the flight deck according to customer requirements on an articulating arm, embedded within the aircraft side panel, or mounted on the side window. The computer module can be installed on the flight deck, EE bay, or other location best suited to the customer requirements.
Hardware Style	The Computer Module and Display Module are proprietary Goodrich designs and produced in the Goodrich facility in Burnsville, MN. The display is connected to the computer via a separate docking station adapter which can be modified to meet specific installation criteria. The maximum allowable distance between the computer and display can be up to 40 feet.

Goodrich Sensors and Integrated Systems	Location: Burnsville, MN
Accessories	The Goodrich EFB has backup battery and wireless options available. The battery is a separate LRU; the wireless connectivity is accomplished within the computer module. In addition, video capability is included within the computer module which can accept up to 8 cameras.
Applications Supported	
Operating System	Microsoft Windows
Electronic Charts	✓ COTS application (Jeppesen, Lufthansa Systems, EAG)
Electronic Checklists	✓ COTS application
Electronic Documents	✓ COTS application
Flight Performance Calculations	✓ COTS application
Flight Planning	✓ COTS application
Video Surveillance	✓ Goodrich Proprietary Video Application: add cameras to existing hardware, cameras available directly from Goodrich Sensor Systems
Weather	✓ COTS application
Maintenance & Logistics	✓ COTS application
Other	✓ COTS applications (e.g., Enhanced Vision System)

Jeppesen	Location: Englewood, CO, USA
Product(s)	Electronic Flight Bag Solutions
Website(s)	<ul><li>www.jeppesen.com</li><li>Product information: Jeppesen EFB solution</li></ul>
FAA Approvals	<ul><li>FAA AEG Operational Suitability Evaluations completed for Jeppesen Class 1 and 2 EFB, Application Manager, eCharts, and Data Management (Seattle ACO).</li><li>FAA AEG Operational Suitability Evaluations completed for Jeppesen eCharts within the Boeing Class 3 EFB system (Seattle ACO).</li><li>FAA AEG Operational Suitability Evaluations and RTCA DO-178B review completed for Airport Moving Map Application approved for the Boeing Class 3 EFB System (Seattle ACO).</li><li>FSB report for Jeppesen Electronic Flight Bag (EFB) Application Software completed, signed, and filed (Seattle ACO).</li><li>FAA AEG Operational Suitability Evaluations completed for Class 1 and 2 document viewer, in conjunction with ARINC Corporation (Seattle ACO).</li><li>Additional Operational Suitability Evaluations underway in support of new functionality for Jeppesen Electronic Flight Bag (Seattle ACO).</li></ul>

Product Overview

Jeppesen offers applications for Class 1, 2, and 3 EFB systems, in partnership with hardware providers. Class 1 systems are portable computing devices, Class 2 systems are mounted solutions, and Class 3 systems are installed equipment solutions. Jeppesen's Class 1 and 2 EFB solutions provide a stable, approved means for controlling applications, data loading, and configuration management. Jeppesen provides its eCharts, Airway Manual Text, and in the future Enroute and Airport Moving Map applications integrated into this EFB environment. These pre-composed and data driven applications provide enhanced situational awareness. In addition, third-parties, such as ARINC, have written applications extending this environment. Jeppesen has teaming arrangements with Astronautics, Rockwell Collins, Goodrich, Teledyne, ARINC, CMC Electronics, and NavAero for hardware, applications, and system integration. Communications functionality is provided through hardware integration (e.g., using ARINC-429, RS-422, ethernet, fiber optic, 802.11 data link (SATCOM, GateLink etc)).

Jeppesen has integrated its navigational application suite as well as data and software management tools with Boeing's Class 3 EFB system. Jeppesen's data management tools integrate with Boeing's TWLU and Communication offering to provide wireless updates to EFBs on the flight deck.

Jeppesen has entered into an agreement to provide its navigational application suite for the Airbus FlySmart system and A380 onboard information system (OIS) as well.

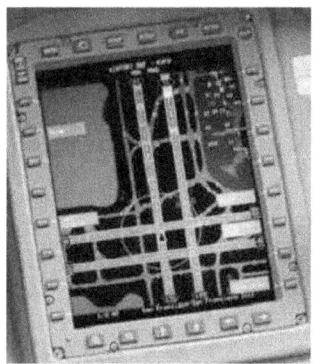
Jeppesen EFB Solution on Boeing Class 3 EFB

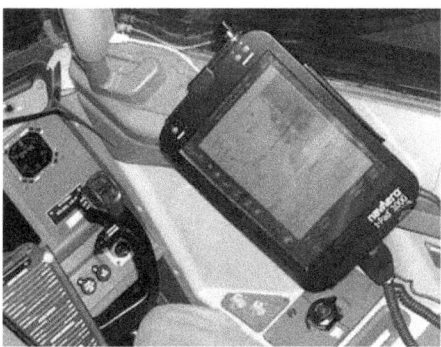
Jeppesen EFB Solution on NavAero t-Bag

| Jeppesen | Location: Englewood, CO, USA |

 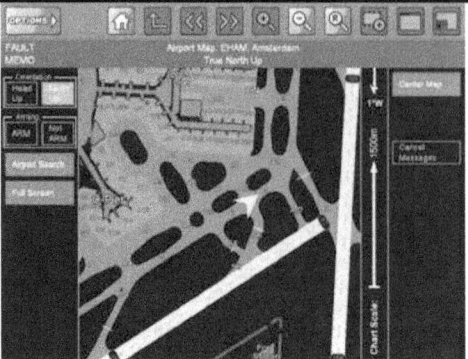

Jeppesen eCharts   Airport Moving Map

Photos courtesy of Jeppesen.

Applications Supported	
Operating System	Varies by system: Microsoft Windows (XP or 2000) and/or Linux are options
Electronic Charts	✓ (Vector-based, pre-composed terminal charts, airport moving maps completed. Data-driven enroute moving map in work)
Electronic Checklists	Possible using COTS application and Jeppesen EFB SDK.
Electronic Documents	✓ (XML and PDF)
Flight Performance Calculations	Possible using COTS application and Jeppesen EFB SDK.
Flight Planning	Jeppesen provides a variety of ground-based flight planning solutions.
Video Surveillance	✓ (Integration with Goodrich and AD Aerospace)
Weather	Possible using COTS application and Jeppesen EFB SDK.
Other	• Approved data and software distribution and management solutions • Ground tools for EFB configuration management • Complete training solutions including Jeppesen EFB Computer-Based Training • 24/7 support

navAero	Location: Chicago, IL
Product(s)	t•Bag C2^2
Website(s)	• www.navaero.com • Product Information: t•Bag C2^2
FAA Approvals	• STCs received on B737-NG (Atlanta ACO, Miami FSDO), MD-10/MD-11 (Atlanta ACO, Memphis FSDO), A319/320/321 (Atlanta ACO, San Francisco FSDO) • STCs in progress A300/310 (Chicago ACO, Memphis FSDO), MD-80 Series (Landmark Aviation DAS, Los Angeles FSDO) • The following related FSB reports have been posted at www.opspecs.com: navAero t Bag(EFB) C2, Jeppesen EFB FSB
Environmental Qualifications	Completed DO-160D testing. Plan to conduct DO-160E testing in April 2007

Product Overview

The t•Bag C2^2 is a Class 2 EFB that consists of a touch screen display with a remote-mounted CPU module and docking station, and an Interface Box that functions as an interconnect point between the CPU assembly and the display. The Interface Box also contains the on/off power switch for the system, a back-up battery indicator lamp and two USB 2.0 ports. navAero maintains strategic partner relationships with COTS providers and systems integrators such as Jeppesen, LIDO, EAG, WingSpeed, Avionica, FlightExplorer, Rockwell Collins, ARINC and others in order to provide customers with a complete hardware and software solution.

Photos courtesy of navAero.

Hardware Class(es)	Class 2
Display Size	10.4" XGA (768 x 1024) or 8.4" SVGA (800 x 600)
Brightness	Optically enhanced for direct sunlight readability. Brightness range from 3 - 750 nits.
Communications	Wireless, ARINC
Controls	• Touch screen • Buttons. 3 hard keys (illumination on/off, brightness increase, brightness decrease).

navAero	Location: Chicago, IL
Mounting and/or Stowage	Designed for side-wall (window frame) peripheral mounting
Hardware Style	Separate display and remote-mounted CPU module
Accessories	ARINC 429 4-channel receiver; GPRS module; CDMA module; WiFi 802.11b/g module
Applications Supported	
Operating System	Microsoft Windows XP or Linux
Electronic Charts	✓ COTS application
Electronic Checklists	✓ COTS application
Electronic Documents	✓ COTS application
Flight Performance Calculations	✓ COTS application
Flight Planning	✓ COTS application
Video Surveillance	✓ COTS application
Weather	✓ COTS application
Other	COTS applications for synthetic vision

OBDS.com (On-Board Data Systems)	**Location:** Mirabel, Quebec
Product(s)	MFB (Multi-Function Flight-Deck Browser), Electronic checklists(ECL), Content Management and Distribution
Website(s)	www.obds.com
FAA Approvals	MFB software and sub-components have been integrated with a number of Class 2 approved systems including CMA-1100, BJAC8720 and well as many Class 1 devices. The checklist application features "tagged" procedures and revision management per AC120-64. MFB is currently in use by corporate and managed fleets as well as Aircraft Manufacturer flight departments.

Product Overview

OBDS is an EFB systems integrator delivering custom Fleet Software and Data Management solutions for Class 1 and Class 2 EFB systems through its MFB ("Multi-function Flight Browser and Task Manager") and OBDSsync (Web Based Content Management) applications.

The MFB "Multi-function Flight Browser and Task Manager" allows fleets operating a variety of hardware devices to operate with a standardized and optimized interface. MFB provides:

- Optimized shell for launching predefined programs and performing updates
- Controlled environment for stability and simplicity
- Centrally managed configurations, can be distributed with data updates
- Update online or by USB/CD - reports delivered to server
- Interactive forms and feedback to server
- Flight document sync

OBDSsync and Web Based Content Management offers:

- Distribution and subscription management for original equipment manufacturer (OEM) content
- Centrally managed configurations, can be distributed with data updates
- Secure differential updates
- Update online or by USB / CD-ROM
- Update logs and flight reports delivered to server
- Web-based configuration management utilities
- Scheduled updates for databases and system updates

Special screen color reversal assists in night operations. Software functionality includes automatic internal brightness management for specific hardware devices and support for communications connectivity (e.g., 802.11 abg, Sprint broadband, GSM etc).

The MFB may be used in a "utility" mode on desktop or laptop PC systems for synchronized library content between office, home and EFB.

OBDS.com (On-Board Data Systems)   Location: Mirabel, Quebec

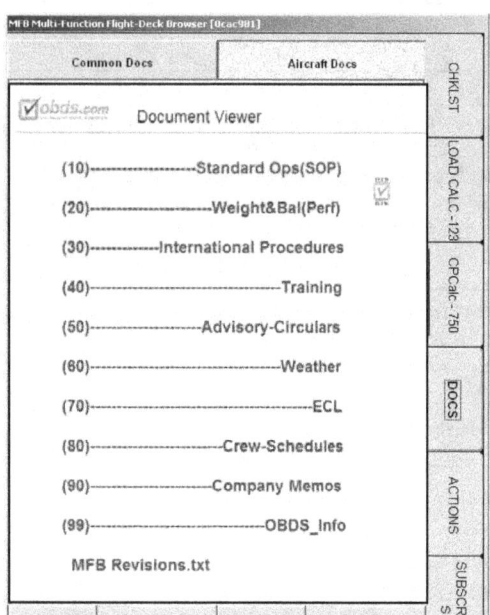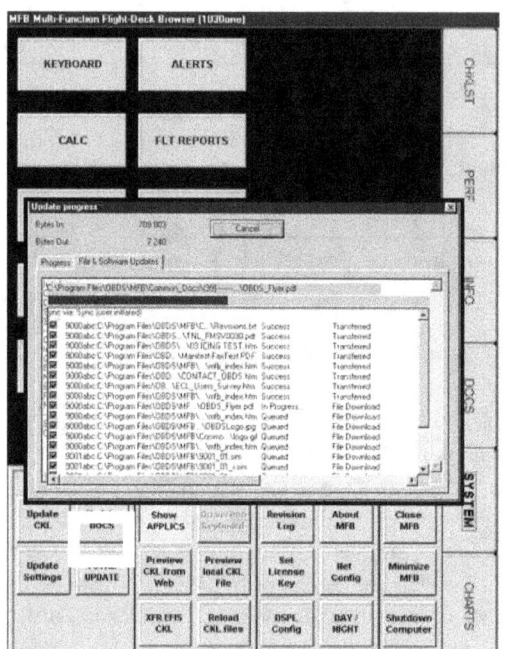

Photos courtesy of OBDS.

Applications Supported

Operating System	Microsoft Windows 2000, XP Pro, XP tablet (Adaptable to all Class 1 and 2 platforms)
Electronic Charts	✓ COTS application. Certain chart systems are made interactive with note-taking capability. Integration for all electronic charting applications. Display and GUI is optimized and customized for controlling third-party charting applications including Jeppesen NACO and EAG.
Electronic Checklists	✓ Class 3 ECL Systems for Honeywell Data Nav, SPZ, Primus 1000, Primus 2000/2000XP, Honeywell Primus EPIC, and Collins E85-86, PL4 among others Error checking and automated error checking functionality available. EPADs – Tabulated Data Formatting and Validation. Tab data incorporated in ECL for fast access and retrieval of Performance data in flight. Class 2 ECL Systems (dedicated EFB checklists) offer custom formatting, open and closed loop ECL, active or passive interactivity, and change review feature (with automated review of changes from last revision)
Electronic Documents	✓ MFB provides a docs viewer with additional features. Note-taking is supported for all documents. Notes may be typed or handwritten, and pilot notes and signatures may be synchronized with other users. Forwarding of reports, and notes to approved recipient list also provided.
Flight Performance Calculations	✓ COTS application. Custom integration supported
Flight Planning	✓ Flight Docs may be delivered to EFB by Online Sync or USB update
Video Surveillance	✓ COTS application. Custom integration supported
Weather	✓ COTS application. Custom integration supported
Other	Integration of COTS applications provided

Paperless Cockpit	Location: Memphis, TN
Product(s)	FliteServ C2-05 EFB with E-Board C2 Remote Wireless Flight Display; E-Board XP3i EFB; Wireless DataLink Server (WDLS)
Website(s)	www.paperlesscockpit.com
FAA Approvals	EFBs are implemented under FAA's guidance regarding Class 1 and Class 2 EFBs, based on the AC 120-76A and the latest Job Aid documents
Environmental Qualifications	**FliteServ C2-R05** is DO-160E qualified. **E-Board C2** carries Eurocad certification that is DO-160E equivalent. All systems are tested via an EMI ground check procedure and an EMI flight check that addresses altitude issues. Other qualifications: **FiteServ** carries the MilSpec certification, **E-Board XP3i** carries FCC, CE and UL certifications, & complies with MilSpec EMI requirements.

Product Overview

Paperless Cockpit's *FliteServ C2-05 EFB* is a military grade two-part EFB system, used with *E-Board C2 Wireless Remote Flight Display*. *E-Board XP3i* is a stand-alone EFB based on the convertible pen tablet computer form factor. All Paperless Cockpit's latest EFBs have been optimized for daytime and nighttime viewing with proprietary screen enhancements. Single-seat and dual-seat configurations are offered. Custom configurations available upon request. *Wireless DataLink Server* is a Portable Electronic Device that allows multiple EFB applications to access GPS data sources at the same time and offers simultaneous access from multiple EFBs to most weather receivers.

FliteServ C2-05 with Wireless Remote E-Board C2 Display

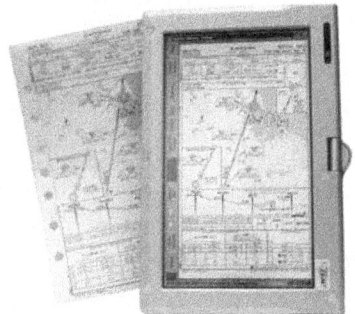

E-Board XP3i

Wireless DataLink Server (WDLS)

Photos courtesy of Paperless Cockpit

Hardware Class(es)	Class 1 and 2
Display Size	**E-Board C2**: 8.4" SVGA (600 x 800) **E-Board XP3i**: 8.9"16:9 XGA (1024 x 600) All Paperless Cockpit displays support rotation to landscape and portrait modes
Brightness	Readable in direct sunlight and dimmable for night flight. Screens optimized for nighttime viewing with several "night vision" software filters offered through proprietary display software.
Communications	**FliteServ C2-05 Processing Unit:** Integrated 802.11G WiFi, built-in 10/100 Ethernet port **E-Board C2 Wireless Remote Display:** Integrated 802.11b WiFi, PCMCIA card slot, compact flash card slot (for wired back up if needed via an Ethernet adapter) **E-Board XP3i EFB:** Built-in 10/100 Ethernet, integrated 802.11b WiFi, on-board Bluetooth version 1.1, built-in 56K modem, W-WAN: 3.5G

Paperless Cockpit	Location: Memphis, TN
	(HDSPA/UMTS/EDGE/GPRS/GSM) **Wireless DataLink Server (WDLS):** 802.11b WiFi, 10/100 Ethernet port
Controls	**E-Board C2 Remote wireless display** • Passive touch screen that provides for data access without a stylus, e.g., using a fingertip • Stylus – passive stylus pen stowed in the slit of the display enclosure • Buttons – on-screen software buttons guide through main functions of the display to communicate with the FliteServ C2-05 Processing unit. • On-screen and USB keyboards available **E-Board XP3i EFB** • Passive touch screen • Stylus – passive stylus pen stowed in the slit of the display enclosure • Buttons – EFB Manager on-screen controls provide easy one-click access to all main software and functionality of the EFB • Mouse/cursor – small on-board track-ball mouse for use in laptop mode • On-screen (for EFB mode) and physical on-board keyboard (for laptop mode) are available
Mounting and/or Stowage	Various mounting solutions including yoke mounts, articulating armatures, side mounts and kneeboards
Hardware Style	**FliteServ C2-05**: Avionics-grade proprietary modular computer processing unit. Low power consumption, EMI-shielded. **E-Board C2**: A remote display for FliteServ family of systems, display is based on the form factor of a mobile .Net device, is a wireless remote display for the FliteServ family of processing units. Compact and lightweight, it offers passive daylight readable touch screen display. **E-Board XP3i**: Based on the form factor for a convertible pen tablet computer, is highly portable, compact, all-in-one EFB.
Accessories	All the necessary accessories for each of Paperless Cockpit's EFBs are included into the individual EFB bundles. Optional accessories most frequently requested include: • Paperless Cockpit's Wireless DataLink Server (WDLS) – allows simultaneous wireless access to in-cockpit weather and GPS data for multiple EFBs in the cockpit • PCMCIA Dual I/O cards – mostly used with E-Board XP3i devices. Offer wired connection to weather and/or GPS sources • Mounts and kneeboards • Spare batteries for the E-Board XP3i EFBs • Additional stylus pens
Applications Supported	
Operating System	Microsoft Windows (XP Professional Edition, standard; Server 2003, optional), Linux
Electronic Charts	✓ (COTS, e.g., JeppView FliteDeck 3.2)
Electronic Checklists	✓ (COTS and custom-built on-demand Paperless Cockpit service)
Electronic Documents	✓ (COTS)
Flight Performance Calculations	✓ (COTS, e.g. UltraNav)
Flight Planning	✓ (COTS, e.g. FlightMap or JeppView)
Video Surveillance	✓ (COTS)
Weather	✓ (COTS, e.g., WSI InFlight, WxWorx on Wings)
Other	Paperless Cockpit EFB Manager and Display Manager software (proprietary). Other supported software applications include Ultra-Nav, MountainScope, and others. Consult Paperless Cockpit's approved software listing (available on the web site) for the full range of supported software.

Rockwell Collins		**Location**: Cedar Rapids, IA
Product(s)	eFlight	
Website(s)	• www.rockwellcollins.com • Product information: eFlight	
FAA Approvals	• eFlight Ground System is installed and in use at more than 40 airlines worldwide • Regulatory approval (STC) attained for multiple eFlight EFB hardware installations. Astronautics Class 3 EFB and the NAT Seattle Network File Server approved as part of the aircraft's Type Certificate. DRS Technologies Class 1 Hammerhead tablet PC; installation STC to be awarded by the Los Angeles Aircraft Certification Office and held for Rockwell Collins by Hollingsead International. • eFlight is currently supporting Operational Approval processes with several regulatory authorities in Europe and the Far East	

Product Overview

What is eFlight?

eFlight is an end-to-end, airline-tailored hardware and software solution that enables effective integration and management of the airline's information flow. The eFlight solution includes decision support applications that address business process optimization of airline operations, aircraft maintenance, and passenger services activities.

eFlight is comprised of 3 core elements:
- eFlight Ground System (EGS), an information management software application for integrating "back office" IT systems (e.g., flight operations, engineering, maintenance, crew scheduling, accounting, etc.)
- eFlight Communications Software (ECS), a software application (air and ground components) for managed, automated, and optimized transfer of data between aircraft and ground. ECS couples guaranteed data delivery with the lowest possible communication costs.
- eFlight Aircraft Software (ACS), applications (eFlight and third party integration) that give flight deck and cabin crews the information they need for effective and efficient performance.

eFlight Functionality

Operating System
  Microsoft Windows

Information management
Flight Operations – Aircraft (ECS/ACS)
- Interactive electronic forms and workflow management: electronic flight folder (EFF), load sheet, technical log, journey log, cabin logs, passenger manifest
- Electronic document viewer, organizer, extended search capability
- Performance calculations – take off and landing (integration of OEM application or optional eFlight version)
- Navigation support – integration of electronic charts and maps, weather application
- Video surveillance integration
- Utilities – Virtual keyboard, electronic signature capability

Flight Operations - Ground (EGS/ECS)
- Departure control/dispatch
- Operations control
- Flight following/aircraft situation display

Cabin and Passenger Management (ECS/ACS)
- Cabin forms
- Duty-free sales
- Passenger affinity tools (special preferences and needs, elite passengers, etc)

| Rockwell Collins | Location: Cedar Rapids, IA |

Ground Information Management (EGS/ECS)
- Enterprise application/ground IT integration
- Remote content management
- Records archiving and configuration control
- Maintenance reporting / technical logs

Hardware
eFlight functions may be hosted on Class 1, 2, or 3 EFB hardware. Each eFlight solution is tailored in accordance with the operator's needs and fleet configuration.

Communications and connectivity
Multi-link communication and communication management (gatelink, satellite, VHF, GSM, USB)

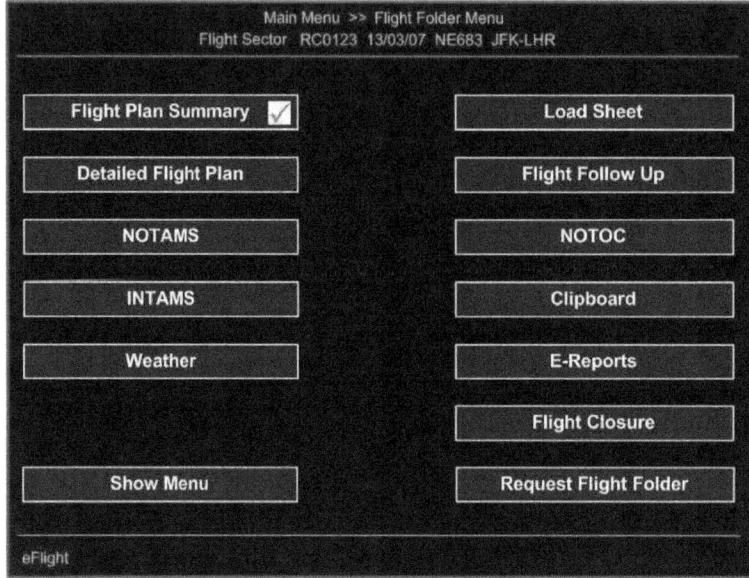

Photo courtesy of Rockwell Collins.

Teledyne Controls	**Location:** Los Angeles, CA
Product(s)	Class 1 EFB, Class 2 EFB, Airbus Class 3 Onboard Information Terminal (OIT), Teledyne EFB Software Suite
Website(s)	<ul><li>www.teledyne-controls.com</li><li>Product information: Aircraft Information Solutions, EFB, hardware options, software applications, Onboard Information Terminal (OIT)</li></ul>
FAA Approvals	Laptop installations include A300/310, B727, DC-10, MD-10 and MD-11. Class 1 certifications include A319/320/321, B737NG, B747, B757, DC-10 (approval for Class 1 docking station obtained through Chicago ACO). Class 2 certifications include A330 and B777, and currently in certification process for the Class 2 Mount and Installation Kits on B737NG and A320 aircraft. Class 3 OIT certified on Airbus Long Range aircraft (except A380).
Environmental Qualifications	Class 1 and 2 EFB systems undergo RTCA DO-160E testing for EMI and decompression. Class 3 systems are RTCA DO-160E qualified.

Product Overview

Teledyne Controls' EFB systems consist of ruggedized computers that host a wide range of software applications. Class 1 EFB systems are tablet computers running Microsoft Windows XP; a certified docking station for storage, a battery charger, and network connectivity to other systems is provided. Class 2 systems consist of two tethered displays and two processor units, mounted in the flight deck. Class 3 systems (e.g., the Airbus Onboard Information Terminal) consist of two-mounted display terminals, each of which are connected to a processor unit in the avionics bay. Cross-talk functionality between the two displays is supported in both Class 2 and Class 3 EFBs.

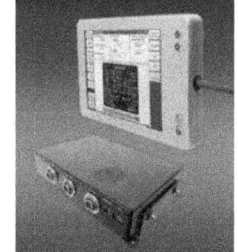

Class 1   Class 2   Class 3

Photos courtesy of Teledyne Controls

Hardware Class(es)	Class 1, 2, and 3
Display Size	8.4" or 10.4" (Class 1 or 2 systems); 12.1" XGA (1024 x 768) (Class 3 system)
Brightness	Displays enhanced for readability in bright sunlight and dimmable for low-light conditions. Class 2: 10.4" display offers brightness range from 7 to 500 Nits; 8.4" display brightness ranges from 7 to 600 Nits Class 3 12.1" and 10.4": 1 to 500 nits, auto-adjusted with manual override
Communications	Class 2: Ethernet, 3 USB 2.0 ports, video and control output Class 3: Ethernet, USB, ARINC 429 Inputs, video and control output.
Controls	<ul><li>All have touch screen interface, and the Class 1 has a tethered stylus.</li><li>Class 2 device has buttons for touch screen, backlight on/off, manual dim and bright. Soft keyboard is also provided.</li><li>Class 3 device has buttons for Display Select (Captain OIT video, First Officer OIT video, LRU1 video & control, LRU2 video & control), CPU Select (Captain CPU, First Officer CPU; Crosscheck; Reset; BRT; DIM), cursor control with mouse left & right clicks and optional USB keyboard with touchpad that connects to the display unit. These buttons perform selected functions required by Airbus software. Optional keyboard integrated via a USB connection available.</li></ul>

Teledyne Controls	**Location:** Los Angeles, CA
Mounting and/or Stowage	Class 1: Portable Class 2: Mounted. The general display unit (DU) location is near the Captain/First Officer side window. It is mounted on an adjustable, quick-release attachment system to allow various users to adjust DU viewing angles and to provide quick removal of the unit. The processor unit (PU) is located in the cockpit and the location varies by aircraft type (Typical locations include coat closets, nooks behind crew seats, or non-structural bulkheads/panels). The PU mount is a quick-release type. Class 3: Installed. 12.1" Display mounted in retractable sliding arm. 10.4" display cockpit side or window mounted. Detachable keyboards mount into sliding arm when in use / detached & stowed when not in use. Processor units located in the forward lower bay. Certified mount/dock provided for Class 1 and 2 installations. Service Bulletin provided by Airbus for Class 3 OIT.
Hardware Style	Class 1 systems are based on COTS laptop and pen tablet computers. Class 2 is a Tethered Display to separate computing platform. Class 3 is a custom built computer that meets Airbus specifications for Long Range and Single Isle aircraft, except for the A380. For both Class 2 and Class 3 systems, the processing box is separate from the display unit. Total distance 6.5 meters from processor unit to display (12.1" or 10.4").
Accessories	Optional keyboard for OIT Class 3 EFB on Airbus installations. Airborne Server Unit, Aircraft Wireless LAN Unit for 802.11 and cellular file transfers, installation kits, mounts and antennas.
Applications Supported	
Operating System	Windows O/S with Flight Manager (Customized application management system; open-systems design providing an integrated operating environment for data sharing between hosted EFB applications, and a suite of administrative utilities pertaining to device configuration (e.g., screen orientation), and content management (e.g., software/data revisions)).
Electronic Charts	✓ Vector-based. Charting application displays most vendors navigation charts Chart Viewer uses an access methodology so pilots can organize and view pdf-based charts from National Aeronautical Charting Office (NACO) and the European Aeronautical Group (EAG). Charts in proprietary formats (e.g., Jeppesen charts) require the chart vendor's application for viewing at this time.
Electronic Checklists	
Electronic Documents	✓ Markup language. Document Viewer is available for various formats, adapting to the documentation technology used by the operator. A search function is included.
Flight Performance Calculations	✓ Weight & Balance, Takeoff and Landing Performance. Computations based on real-time airport weather, aircraft-specific data, and MEL/CDL conditions. Compatible with digital AFM from aircraft OEMs. AEG approved. Several aircraft types already in service.
Flight Planning	
Video Surveillance	✓ Cabin Surveillance Viewer (optional). Compatible with a variety of surveillance system suppliers on the market. Depending on the capabilities of the chosen camera system, a range of real-time viewing and playback options is available.
Weather	
Other	SmartForms that allow flight crews to complete forms using the EFB touch-screen interface, and ground based content and configuration management workstation software.

Universal Avionics	Location: Tuscon, AZ
Product(s)	Application Server Unit (ASU) / Displays on Universal Cockpit Display Terminal (UCDT-III) or EFI-890R
Website(s)	- www.uasc.com - Product Information: ASU
FAA Approvals	- TSO C113 and TSO C165 for internally-developed electronic approach charts - STC'd on a wide range of aircraft including Falcon 10, 20, and 50, King Air 200/300350, Pilatus PC-12, Boeing Business Jet and 737-300/400, Bombardier's Lear 25, Challenger and Global Express, Dassault Falcon 2000, Gulfstream G500, and Cessna Citation Bravo. - All product approvals (TSO) and those installation approvals (STC) obtained by Universal Avionics are issued from the Los Angeles ACO - Certified with Level C software (DO-178B)
Environmental Qualifications	There are two installation options: (1) if integrated with the instrument panel, then all components are DO-160E qualified; (2) if the display unit is separate (e.g., UCDT-III + ASU), then the components are a mix of RTCA DO-160D and RTCA DO-160E qualified.

Product Overview

The ASU provides a supplemental electronic display system that can be integrated with flight deck instruments. The remote ASU computer supports up to two display terminals (UCDT-III) or can be displayed on the EFI-890R Navigation Display (ND). Pilots can access electronic charts, checklists, electronic documents, WSI satellite weather and video sources. The ASU electronic chart database is provided by Jeppesen's JeppView product; appropriate charts can be sorted automatically based on departure and arrival airport information supplied by the Flight Management System (FMS). Aircraft present position can be displayed on the electronic charts or WSI broadcast weather products for all phases of flight. Aircraft-specific procedural checklists can be created by pilots for normal, abnormal, and emergency situations. Documents such as Flight Manuals can be digitized and stored for convenient in-flight access.

Photos Courtesy of Universal Avionics

JeppView Chart© reproduced with the permission of Jeppesen Sanderson, Inc. NOT FOR NAVIGATIONAL USE.©Jeppesen Sanderson, Inc. 2007

Universal Avionics	Location: Tuscon, AZ
Hardware Class(es)	Class 3
Display Size	UCDT-III: 8.4" SVGA (1024x768) EFI-890R ND: VGA (780x780)
Brightness	Low reflectance, fully dimmable for nighttime viewing
Communications	Ethernet, RS-232, RS-422, and ARINC 429
Controls	UCDT-III has touch screen EFI-890R ND has cursor control panel
Mounting and/or Stowage	Cockpit or yoke-mountable
Hardware Style	Custom built hardware, with separate display and processor units. Connectivity with the UCDT-III is via LVDS (low voltage differential signaling); the EFI-890R interface utilizes VGA
Accessories	COTS HP (Hewlett-Packard) printer for onboard chart, and electronic document printing.
Applications Supported	
Operating System	Embedded Microsoft Windows XP
Electronic Charts	✓ Vector-based; Jeppesen electronic charts with worldwide coverage and printing.
Electronic Checklists	✓ Viewer only. Checklists are interactive and interface to yoke switches. Aircraft specific checklists are supported. Offline program to create aircraft unique checklist is provided by Universal.
Electronic Documents	✓ Viewer only: PDF files can be loaded, viewed, deleted, printed and managed within the viewer application.
Flight Performance Calculations	
Flight Planning	
Video Surveillance	✓ Support for RS-170 (e.g. EVS camera) and NTSC (e.g. standard camera) video inputs.
Weather	✓ WSI Embedded application viewer
Other	HP450/460 printer support. Other HP printers can be tested for compatibility.

Virtual Papyrus Inc.	**Location:** Sun Valley, CA
Product(s)	MCP System (Mobile Computing Platform) as part of the Class II Plus Offering consisting of: • ANKH (Aircraft Network Konnectivity Hardware) main computing platform with multiple CPUs • RAH T1X (Remote Access Hardware) series designed for Boeing, Bombardier, Embraer, ATR, and similar space restricted cockpits • RAH T3X (Remote Access Hardware) series designed specifically for Airbus Flight Decks • Support Accessories (additional displays, battery packs etc)
Website(s)	www.virtualpapyrus.com
FAA Approvals	Complies with FAA AC120-76A as a Class 2 system
Environmental Qualifications	Presently undergoing qualifications testing for RTCA DO-160E

Product Overview

The Virtual Papyrus MCP system is designed by airline professionals to meet airline operational requirements. Its main strengths are:
- Communication through multiple paths to meet an airlines logistic issues
- Computing power to drive the flight deck and the passenger cabin
- Flexibility to adapt to the ever changing PC hardware upgrade path by simple sub-component swap capability
- Flexibility to run any Windows based application the airline requires
- And the grace of integral cockpit design for ergonomic compliance

Photos courtesy of Virtual Papyrus, Inc.

Hardware Class(es)	The MCP is a Class 2 system. It utilizes a patent pending design which combines the strength of a split architecture, screen from a CPU, COTS hardware components, and multiple communications paths for logistical support.
Display Size	RAH units are based on COTS LCD Technology. • RAH T1X Series = 9" - 16x9 widescreen format @ WVGA • RAH T3X Series = 12.1" - 16x9 widescreen format @ WXGA
Brightness	All the RAH series use Superbright LCD technology for daylight readability and are all incrementally step dimmable to zero for display control at night. They also support a 160 degree viewing angle.
Communications	Communications is the strength of the MCP System providing connectivity via: telephony (GPRS,GSM,1X,EVDO,EDGE, HSDPA covering both world present standards and the new HSUPA world wide high speed network technology), 802.11 WIFI (A,B,G), 8 port Ethernet Hub, 5 USB interfaces, ARINC 429, RS 232 & RS 422 & RS 485
Controls	The RAH series has the following common elements: • Touch Screen: self calibrating throughout operating temperature range.

Virtual Papyrus Inc.	Location: Sun Valley, CA
	Separate toggle control provided to adjust screen brightness (full bright to 0) • Integrated indestructible keyboard, spill proof, chemical resistant, and sealed with 5 stages of integral back lighting (0 to high) for night or day use. 10,000,000 keystrokes MTBF. • USB ports. These may be operator controlled at the airlines security level for activation or deactivation as required. The ANKH component is self monitoring with individual CPU health status displays. It implements auto update and auto backup capabilities.
Mounting and/or Stowage	The RAH units are mounted with quick change capabilities for a modular design. The RAH units have different mount locations (e.g., the sliding window, the built in chair table) and custom installations as per individual requests. The ANKH is a mounted housing for the industrial COTS laptop, CPU boards, and power supplies.
Hardware Style	• The RAH series of screens/keyboards are COTS units, repackaged by Virtual Papyrus as part of the mounting solution. The RAH contains only the screen, keyboard, USB, and brightness controls. • The ANKH houses the CPU boards, communications and power supplies. The CPU boards are COTS Intel based dual-core mobile processor units repackaged by Virtual Papyrus. Each board is individually changeable like a laptop computer. • There is no distance restriction between the ANKH and the RAH as this connection is achieved by fiber optics. The placement of the ANKH within the fuselage is usually more the consideration of maintenance personnel for convenient access. This design ensures the flight deck remains intact as designed by the OEM, with no additional components to take up existing allocated space.
Accessories	• Depending on the level of software applications the MCP is hosting, the Airline may elect to procure the optional battery pack. • Virtual Papyrus offers super low light IP cameras for cabin surveillance.

Applications Supported

Virtual Papyrus embraces the total open architecture concept and therefore offers a variety of software products in conjunction with our business partners. Virtual Papyrus does not limit the Airline to proprietary applications.

Operating System	Microsoft Windows™ XP (Linux will be considered for future installation upgrades)
Electronic Charts	✓ COTS raster and vector-based charting applications (e.g., Blueskydox, FlightPrep, Jeppesen, Lido, Maptech, EAG NAVTECH)
Electronic Checklists	✓ COTS applications
Electronic Documents	✓ Custom and COTS applications. Functionality supported includes: • **Viewer with Additional Features**: The base system, in PDF format, has the added benefit of multi-document hyperlinks and sophisticated search criteria. Additional standard features include incremental updating via electronic media. • **Markup Language**: Virtual Papyrus's Document Management Business Partner offers a full suite of aviation related documentation in XML language including Flight Operations and Maintenance Data.
Flight Performance Calculations	✓ COTS applications (The Virtual Papyrus MCP can support both Type A and Type B Software Flight Performance applications)
Flight Planning	✓ COTS applications (Can host cockpit centric flight planning systems like FlightPrep, etc)
Video Surveillance	The MCP supports IP cameras via the 8 port Ethernet hub. These can be daisy-chained in multiple of eights to a max of 256. Most installations require 2-4 cameras. The MCP can interface with analog cameras; however an additional analog/digital conversion hardware piece is required.
Weather	✓ COTS applications (Honeywell (WINN),WxWorx, Rockwell, WSI, and Jeppesen)
Other	The MCP has the functionality and communication suite to run the Airbus E-Logbook application and similar applications.

Software Providers

The following table provides a list of EFB software providers. The most common applications for EFBs are electronic charts, electronic checklists, electronic documents, flight performance calculations, flight planning, cabin video surveillance, and weather. Other functionality may also be provided, such as the following:

- Note-taking applications that allow pilots to write notes and save them
- Logbooks
- Enhanced vision systems/synthetic vision systems
- Electronic Mail
- Surface moving map
- CDTI traffic displays
- Terrain awareness systems
- Forms

In the table below, the name of the software provider is hyperlinked to that company's website. More information on the products and services offered can be found by selecting the name of the software provider.

PROVIDER	OPERATING SYSTEM			APPLICATIONS							
	Microsoft Windows	Linux	Other	Electronic Charts	Electronic Checklists	Electronic Documents	Flight Perf. Calc.	Flight Planning	Video Surveillance	Weather	Other
1. Adobe	✓					✓					
2. Advanced Data Research (ADR)	✓		✓	✓	✓						✓
3. Aviation Communication & Surveillance Systems (ACSS)			✓								✓
4. Aero Data Solutions						✓					
5. Airbus											
6. Aircraft Data Fusion	Platform-independent			✓	✓	✓	✓		✓	✓	✓
7. Aircraft Management Technologies (AMT)			✓	✓	✓	✓	✓	✓	✓		✓
8. Aircraft Performance Group	✓					✓					
9. AirGator	✓			✓	✓	✓	✓	✓			✓
10. Astoria Software						✓					
11. Astronautics	✓	✓							✓	✓	✓
12. CMC Electronics	✓		✓		✓	✓	✓	✓	✓		✓

PROVIDER	OPERATING SYSTEM			APPLICATIONS							
	Microsoft Windows	Linux	Other	Electronic Charts	Electronic Checklists	Electronic Documents	Flight Perf. Calc.	Flight Planning	Video Surveillance	Weather	Other
13. Control Vision	✓			✓	✓		✓	✓			✓
14. Flight Deck Resources	✓	✓		✓		✓					✓
15. Flight Explorer	✓							✓			
16. Flightprep (Stenbock & Everson)	✓			✓	✓		✓	✓			✓
17. Hangar B-17	✓							✓			
18. Honeywell WINN			✓	✓						✓	
19. ION Systems	✓					✓					
20. Jeppesen	✓	✓	✓	✓	✓	✓	✓	✓		✓	
21. Lido				✓		✓	✓	✓			
22. Maestro Aviation Limited					✓		✓				
23. Maptech	✓			✓							
24. MyAirplane.com	✓			✓							
25. Navtech	✓			✓							
26. On-Board Data Systems (OBDS)	✓		✓		✓						
27. RMS Technology	✓						✓				
28. Teledyne Controls			✓	✓		✓	✓		✓		
29. True Flight	✓			✓				✓			
30. Ultramain efbFlightLogs	✓										✓
31. Ultra-Nav	✓							✓			
32. Virtual Papyrus Blueskydox	✓			✓							
33. WSI	✓									✓	
34. WxWorx	✓									✓	

References

EFB Policy Documents

Federal Aviation Administration, Advisory Circular AC 120-76A. *Guidelines for the certification, airworthiness, and operational approval of electronic flight bag computing devices.* (17 March 2003) Available at www.faa.gov, under Regulations and Policies.

Federal Aviation Administration, Notice 8200.98, Electronic Flight Bag Job Aid. (13 October 2006). Available at www.faa.gov, under Regulations and Policies.

Joint Aviation Authorities (JAA). Temporary Guidance Leaflet No. 36. Approval of Electronic Flight Bags (EFBs). (June 2004) Available at www.jaa.nl/secured/Operations/Public%20Documents/TGLs/TGL%2036.pdf.

Transport Canada, Commercial and Business Aviation Advisory Circulars (CBAAC) No. 0231. Electronic Flight Bags (20 July 2004). Available at www.tc.gc.ca/civilaviation/commerce/circulars/AC0231.htm.

RTCA, Inc. Documents

RTCA/DO-160E, *Environmental Conditions and Test Procedures for Airborne Equipment.*

RTCA DO-178B, *Software Considerations in Airborne Systems and Equipment Certification.*

FSB Reports (as of April 2007)

The following Final FSB reports are available at www.opspecs.com/AFSData/FSBRs/Final/:

Arinc Messenger (NavAero), Class 2 Electronic Flight Bag (EFB), ST3064DE-T. Available at www.opspecs.com/AFSData/FSBRs/Final/B-727%20EFB%20FSB/

Flight Options, Fujitsu Stylistic LT C-500, Class 2 Electronic Flight Bag (EFB); Noga Engineering model "HOLD IT model MG" Mounting System. Available at www.opspecs.com/AFSData/FSBRs/Final/BE-400A%20Fujitsu%20EFB/

Boeing Class 3 Electronic Flight Bag – Block Point 3 (Operational Suitability Report). Available at www.opspecs.com/AFSData/FSBRs/Final/Boeing%20Class%203%20EFB%20FSB/

Advanced Data Research, FG-3600, FG-5000, Class 2 Electronic Flight Bag (EFB), EMB-135BJ (Legacy); Mounting System: Audio International Articulating Arm. Available at www.opspecs.com/AFSData/FSBRs/Final/EMB-135%20EFB%20FSB/

Fujitsu LT P-600, Class 1 Electronic Flight Bag (EFB). Available at www.opspecs.com/AFSData/FSBRs/Final/Fujitsu%20LT%20P-600%20EFB%20FSB/

Jeppesen, Electronic Flight Bag (EFB) Application Software. Available at www.opspecs.com/AFSData/FSBRs/Final/Jepp%20EFB%20FSB/

navAero t•Bag C22, Class 2 Electronic Flight Bag (EFB); Mounting System: Avionics Support Group (ASG®). Available at www.opspecs.com/AFSData/FSBRs/Final/navAero%20t%20bag%20(EFB)/

Xplore Technologies' iX104C2, Class 1 Electronic Flight Bag (EFB). Available at www.opspecs.com/AFSData/FSBRs/Final/Xplore%20iX%20%20104%20C2%20EFB%20FSB/

Draft FSB reports are also available at www.opspecs.com/AFSData/FSBRs/Draft/.

Volpe Center EFB Research Reports

The following documents are available at www.volpe.dot.gov/hf/aviation/efb.

Chandra, D. C. and Yeh, M. (2006). *A Tool Kit for Evaluating Electronic Flight Bags*. DOT/FAA/AR-06/44. DOT-VNTSC-FAA-06-21. Washington, DC. U.S. Department of Transportation, Federal Aviation Administration.

Yeh M. and Chandra, D. C. (2005). *Electronic Flight Bag (EFB): 2005 Industry Review*. DOT-VNTSC-FAA-05-06, USDOT Volpe Center: Cambridge, MA.

Chandra, D. C., Yeh M., & Riley, V. (2004). *Designing a Tool to Assess the Usability of Electronic Flight Bags (EFBs)*. DOT/FAA/AR-04/38, USDOT Volpe Center: Cambridge, MA.

Chandra, D. C., Yeh M., Riley, V., & Mangold, S.J. (2003). *Human factors considerations in the design and evaluation of Electronic Flight Bags (EFBs), Version 2*. DOT-VNTSC-FAA-03-07. USDOT Volpe Center: Cambridge, MA.

Chandra, D. C. and Mangold S. J. (2000). *Human factors considerations in the design and evaluation of electronic flight bags (EFBs) Version 1: Basic functions*. DOT-VNTSC-FAA-00-22. Cambridge, MA: USDOT Volpe Center.

Volpe Center EFB Conference Papers

The following short research papers, presented at various conferences, are also available at www.volpe.dot.gov/hf/aviation/efb.

Chandra, D.C. and Yeh, M. (2006) Evaluating Electronic Flight Bags in the Real World. *Proceedings of the International Conference on Human-Computer Interaction in Aeronautics (HCI–Aero) 2006*. 20–22 September 2006, Seattle, Washington.

Chandra, D.C. and Yeh, M. (2004). Designing and Testing a Tool for Evaluating Electronic Flight Bags. *Proceedings of the International Conference on Human-Computer Interaction in Aeronautics (HCI–Aero) 2004*. 29 September – 1 October 2004, Toulouse, France.

Chandra, D.C. (2003). A tool for structured evaluation of electronic flight bag usability. In *Proceedings of the 22nd Digital Avionics Systems Conference (DASC)*. 12–16 October 2003, Indianapolis, IN.

Chandra, D. C. (2002). *Human Factors Evaluation of Electronic Flight Bags*. Proceedings of HCI-Aero 2002. 23-25 October. Cambridge, MA.

Chandra, D. C. & Mangold S. J. (2000) *Human factors considerations for the design and evaluation of Electronic Flight Bags*. Proceedings of the 19th Digital Avionics Systems Conference. 10-12 October 2000, Philadelphia, PA.

Other

Flight Safety Foundation Editorial Staff, "Paperless Cockpit Promises Advances in Safety, Efficiency." Flight Safety Digest Volume 24 (June 2005).

Trademark Notices

The following trademark information was gathered from speaking with company representatives and by searching the database on the United States Patent and Trademark Office website. Other product and company names mentioned herein may be the trademarks of their respective owners. The accuracy of this information can not be guaranteed. The Volpe Center disclaims liability for errors, omissions or future changes.

Adobe is the registered trademarks of Adobe Systems, Inc.

Flightman is the trademark of Aircraft Management Technologies.

Airbus is the registered trademark of Deutsche Airbus GMBH.

Boeing is the registered trademarks of Boeing Company.

CMC Electronics and *PilotView* are the registered trademarks of CMC Electronics Inc.

SkyTab, *FliControl*, *FliPrep*, and *FliView* are the trademarks of Flight Deck Resources.

Jeppesen, *FliteStar*, *Jeppesen NavSuite,* and *JeppView* are the registered trademarks of Jeppesen Sanderson, Inc.

Linux is the registered trademark of Linus Torvalds.

Microsoft, *Microsoft Windows*, *Windows 2000*, *Windows XP, Windows XP Tablet Edition* are the registered trademark of Microsoft Corporation.

Motion Computing is the registered trademark of Motion Computing, Inc.

R-A-M is the registered trademark of National Products, Inc.

MFB and *OBDSsync* are trademarks of OBDS (On-Board Data Systems).

NavAero, t●Bag, and *t●Pad* are the registered trademarks of Navaero AB.

UNIX is the registered trademark of The Open Group.

Paperless Cockpit, E-Board, and *FliteServ* are trademarks of Paperless Cockpit, Inc.

Rockwell Collins and *eFlight* are the registered trademark of Rockwell International Corporation.

ChartCase Professional is the registered trademarks of Stenbock & Everson Inc.

AvVantage is the registered trademark of Teledyne Controls, Inc.

Teledyne is the registered trademark of Teledyne Technologies, Inc.

Universal is the registered trademark of Universal Avionics Systems Corporation.

WxWorx and *WxWorx on Wings* are the registered trademark of Baron Services, Inc.

WSI and *WSI InFlight* is the registered trademark of WSI Corporation.

Websites

The following is a list of websites for EFB system manufacturers and software providers discussed in the industry review. This list was compiled in February, 2007.

System Providers

EFB Systems Manufacturer	Website
Advanced Data Research (ADR)	www.adrsoft.com
Airbus	www.airbus.com
Aircraft Management Technologies (AMT)	www.flightman.com
AirGator, Inc	www.airgator.com
ARINC	www.arinc.com
Astronautics	www.astronautics.com
CMC Electronics	www.cmcelectronics.ca
DAC International	www.dacint.com
Flight Deck Resources	www.flightdeckresources.com
FlightPrep, Stenbock and Everson, Inc.	www.flightprep.com
Goodrich Sensors and Integrated Systems	www.goodrich.com Additional provider: Business Jet Aircraft Completions, www.bjacaerospace.com
Jeppesen	www.jeppesen.com
NavAero	www.navaero.com
On-Board Data Systems (OBDS.com)	www.obds.com
Paperless Cockpit	www.paperlesscockpit.com
Rockwell Collins	www.rockwellcollins.com
Teledyne Controls	www.teledynecontrols.com
Universal Avionics	www.universalavionics.com
Virtual Papyrus	www.virtualpapyrus.com

Software Providers

Software/Content Provider	Website
Adobe	www.adobe.com
Advanced Data Research (ADR)	www.adrsoft.com
Aviation Communication & Surveillance Systems (ACSS)	www.acssonboard.com
Aero Data Solutions	www.aerodatasolutions.com
Airbus	www.airbus.com
Aircraft Data Fusion	www.aircraftdatafusion.com
Aircraft Management Technologies (AMT)	www.flightman.com
Aircraft Performance Group	www.acftperfgrp.com
AirGator	www.airgator.com
Astoria Software	www.astoriasoftware.com
Astronautics	www.astronautics.com

Software/Content Provider	Website
CMC Electronics	www.cmcelectronics.ca
Control Vision	www.anywheremap.com
Flight Deck Resources	www.flightdeckresources.com
Flight Explorer	www.flightexplorer.com
Flightprep (Stenbock & Everson)	www.flightprep.com
Hangar B-17	www.hangarb17.com
Honeywell	www.honeywell.com
ION Systems	www.ionsystems.com
Jeppesen	www.jeppesen.com
Lido	www.lhsystems.com/en/index.htm
Maestro Aviation Limited	www.maestro-aviation.com
Maptech	www.maptech.com
MyAirplane.com	www.myairplane.com
Navtech	www.navtechinc.com
On Board Data Systems	www.obds.com
RMS Technology	www.rmstek.com
Teledyne Controls	www.teledynecontrols.com
True Flight	www.trueflight.org
Ultramain efbFlightLogs	www.ultramain.com
Ultra-Nav	www.ultranav.com
Virtual Papyrus	www.virtualpapyrus.com
WSI	www.wsi.com
WxWorx	www.wxworx.com

www.ingramcontent.com/pod-product-compliance
Lightning Source LLC
Chambersburg PA
CBHW081900170526
45167CB00007B/3087